CAMBRIDGE LIBRARY COLLECTION

Books of enduring scholarly value

History

The books reissued in this series include accounts of historical events and movements by eye-witnesses and contemporaries, as well as landmark studies that assembled significant source materials or developed new historiographical methods. The series includes work in social, political and military history on a wide range of periods and regions, giving modern scholars ready access to influential publications of the past.

Sketches of the Royal Society and Royal Society Club

Sir John Barrow (1764–1848) was a distinguished British government diplomat whose career took him to China and Africa, and who in forty years as Secretary to the Admiralty was responsible for promoting Arctic and Antarctic exploration. A close friend of Sir Joseph Banks, he served on the Council of the Royal Society and as President of the Royal Geographical Society. *Sketches of the Royal Society and Royal Society Club* was published posthumously in 1849, as a supplement to Barrow's autobiography (also published in this series). It consists of a brief history of the societies, followed by a series of memoirs of presidents of the Royal Society of Barrow's time, and of other leading members of the Society and the Royal Society Club, the elite dining club associated with it. The biographies provide abundant evidence of the central importance of the Royal Society to scientific life in nineteenth-century Britain.

Sketches of the Royal Society and Royal Society Club

Sir John Barrow

CAMBRIDGE UNIVERSITY PRESS

Cambridge, New York, Melbourne, Madrid, Cape Town, Singapore,
São Paolo, Delhi, Dubai, Tokyo, Mexico City

Published in the United States of America by Cambridge University Press, New York

www.cambridge.org
Information on this title: www.cambridge.org/9781108028165

This edition first published 1849
This digitally printed version 2010

ISBN 978-1-108-02816-5 Paperback

SKETCHES

OF

THE ROYAL SOCIETY

AND

ROYAL SOCIETY CLUB.

SKETCHES

OF

THE ROYAL SOCIETY

AND

ROYAL SOCIETY CLUB.

BY

SIR JOHN BARROW, BART., F.R.S.

LONDON:

JOHN MURRAY, ALBEMARLE STREET.

1849.

London : Printed by WILLIAM CLOWES and SONS, Stamford Street.

PREFACE.

IT is stated, in a note prefixed to the manuscript of the following pages, that they were intended, by Sir John Barrow, as a Supplementary Chapter to his Autobiography; and were occasioned by a good-natured reproach from a friend, an old member of the Royal Society Club, for his having omitted in his Memoirs all mention of that Club, with which he had been intimately associated full forty years, and of the many learned and agreeable members with whom he had enjoyed so long, and so friendly, an intercourse.

In order to repair what he admitted to be an omission, he threw together, from his own observations and recollections, as well as from several biographies* already published, the following sketches of the six Presidents of the Royal Society, who filled the chair in his time, and of some other distinguished members of the Society and the Club.

In a large portion of such a work there can be, of course, little novelty; but Sir John Barrow thought

* Particularly those by Lord Brougham, Dr. Paris, &c.

that, besides being an appropriate Supplement to his own life (with which object it is printed uniformly with the former volume), it might also perhaps be acceptable in a separate shape, as a slight and compendious account of the Royal Society and of some of its most distinguished members, for the last forty years—not now to be found in any *single* publication.

CONTENTS.

equal to the Park by where
in Gloucester? I join the
early gay Promenade at
nothing very afternoon about
3 o'clock and Dowager in
my carriage backwards and
forwards along the banks I
the aggregated through the
constant Promenaders, Ruby
Hilland, and think I at we enjoy
many more for the most
dangerously indulged
almost all my old friends
and acquaintance one dead,
or something like it, to the
common become still and
very dull and stupid?
very thin, but remembering

Mr. Brande, my dear
father must have applied
to himself: he used often to
say, How many valuable
and old friends have I sur-
vived! How much do I regret
the loss of them.

This sadly was
the last letter I received
from my dear father; it was
about the commencement of
his last illness — water in
the chest. He died on
the 10th March 1847, sensible
to the last.

And I who
survive

survive to inherit his
late Station, the Father
of the Royal Society Club,
am writing this on
the 29th day of October
1848, in the Eighty-
fifth Year of my Age

SKETCHES

OF

THE ROYAL SOCIETY

AND

ROYAL SOCIETY CLUB.

SECTION I.

The Royal Society, its Presidents, and the Royal Society Club.

SOME time after the Civil Wars, a few learned members of the University of Oxford were in the habit of meeting, at stated times, at the lodgings of the celebrated Dr. John Wilkins, of Wadham College, afterwards Bishop of Chester, for literary and scientific conversation. The principal attenders besides Dr. Wilkins were Dr. Seth Ward, afterwards Bishop of Salisbury, Mr. Boyle, Sir William Petty, Mr. Mather Wren, Sir Christopher Wren, with several other ingenious but less remarkable men. In time, of course, many of these were called away from Oxford, on the various businesses of life; but the greater part meeting in the common centre — London, they, about 1658, renewed those meetings at Gresham College, on the

B

Wednesdays and Thursdays, when Sir Christopher
Wren and Mr. Rook delivered their usual Gresham
Lectures. They were soon joined by Lords Brouncker
and Brereton, Mr. Evelyn, Dr. Crone, and many other
literary and scientific men. They were, for the most
part, of royalist politics; and Harrington, in his 'Prerogative of Popular Government,' sneered at them as a
set of *virtuosi*, "who had an excellent faculty of magnify-
ing a flea and diminishing a commonwealth; and on the
Restoration, received an important accession of several
of the eminent persons who had returned with the King
from exile. Amongst these, the most remarkable, or,
at least, the most influential, was Sir Robert Moray,
the descendant of an ancient family in the Highlands
of Scotland, well educated, partly at St. Andrew's and
partly in France; he had been in the army of Louis
XIII, was raised to the rank of colonel, and on the
Restoration was appointed a privy councillor for Scotland. Sir Robert brought this association to the knowledge of the restored King, who warmly approved of it,
and on the 5th December, 1660, Sir Robert communicated to the meeting His Majesty's intention of becoming
their patron. Early in the next year the King declared
himself their patron in form, and gave them the appel-
lation of "The Royal Society;" and on the 6th of
March, Sir Robert Moray was elected President of the
institution. For a couple of years the elections were
monthly, and Sir Robert and Dr. Wilkins were alternately elected. The new Society soon obtained such
consistency, and its objects were so generally approved,
that, towards the close of 1662, the King granted it a
charter of incorporation, prescribed for its maintenance

and government certain statutes, and nominated William, Lord Viscount Brouncker, First President of the Royal Society, with a Council of twenty-one members, a Treasurer, and two Secretaries; all to be renewed annually on St. Andrew's Day, the 30th November.

From that time to the election at the close of the present year, 1848, there have been twenty-six Presidents, of whom the following is a list, with the dates of their election and the duration of their service:—

PRESIDENTS OF THE ROYAL SOCIETY.

	Elected.	Duration.
William, Viscount Brouncker	April 22, 1663	14 years.
Sir Joseph Williamson, Knt.	Nov. 30, 1677	3 years.
Sir Christopher Wren, Knt.	Nov. 30, 1680	2 years.
Sir John Hoskins, Bart.	Nov. 30, 1682	1 year.
Sir Cyril Wyche, Bart.	Nov. 30, 1683	1 year.
Samuel Pepys, Esq.	Dec. 1, 1684	1 year.
John, Earl of Carbery	Nov. 30, 1686	3 years.
Thomas, Earl of Pembroke	Nov. 30, 1689	1 year.
Sir Robert Southwell, Knt.	Dec. 1, 1690	5 years.
Charles, Earl of Halifax	Nov. 30, 1695	3 years.
John, Lord Somers	Nov. 30, 1698	5 years.
Sir Isaac Newton	Nov. 30, 1703	24 years.
Sir Hans Sloane, Bart.	Nov. 30, 1727	14 years.
Martin Folkes, Esq.	Nov. 30, 1741	11 years.
George, Earl of Macclesfield	Nov. 30, 1752	12 years.
James, Earl of Morton	Nov. 30, 1764	4 years.
James Burrow, Esq., ad interim	Oct. 27, 1768	
James West, Esq.	Nov. 30, 1768	4 years.
James Burrow, Esq., ad interim	July 7, 1772	
Sir John Pringle, Bart.	Nov. 30, 1772	6 years.

	Elected.	Duration.
Sir Joseph Banks, Bart.	Nov. 30, 1778...	41 years.
W. Hyde Wollaston, M.D., ad interim .... }	....June 29, 1820	
Sir Humphry Davy, Bart.	Nov. 30, 1820...	7 years.
Davies Gilbert, Esq., M.P., ad interim .... }	.. Nov. 13, 1827	
Davies Gilbert, Esq.	Nov. 6, 1827...	3 years.
H. R. H. The Duke of Sussex...	Nov. 30, 1830...	8 years.
Marquis of Northampton	Nov. 30, 1838.. } Notice of Resignation.....Nov. 30, 1848.. }	10 years.

Of these six last Presidents, whose colleague I
had the honour to be, a few brief notices will be
given hereafter. At present I must speak of myself,
as an autobiographer is entitled to do. I was elected
a fellow of the Royal Society in 1806. Having then
but a narrow acquaintance in London, and being re-
lieved, on the change of Administration, from my official
duties at the Admiralty, I was glad to avail myself of
this introduction into scientific and literary company, for
which I always had a taste, and now had leisure. But
I very soon found myself disappointed in the expectations I had formed of the Royal Society—that is to say,
of much amusement or instruction from the mode in
which the public business at the weekly meetings was
conducted.

The general affairs of the Society were administered
by a President, with a Council, two Secretaries, besides
a Foreign and Assistant Secretary, a Treasurer, and a
Clerk or two. A meeting was held one evening in the
week at eight o'clock, when the President took the
chair. One secretary read an abstract of such paper,

or papers, as had been read on the preceding evening, and the other secretary read a fresh paper or
papers, if any might have come in during the week.
The subjects were generally scientific; and I confess
I often found them dull enough in themselves, and
not always improved by the monotony of an official
reader, sometimes weary of his task.

These defects are satirically alluded to in the ' Pursuits of Literature :'—

> " While o'er the bulk of these transacted deeds
> Prim Blagden pants, and damns them as he reads."

Dr. (afterwards Sir Charles) Blagden was one of the
secretaries.

If the names of any candidates had been suspended
in the meeting-room the required number of weeks,
an election by ballot took place. The ballot-box was
taken to the President, who counted the balls for and
against, and pronounced the result, then making his
bow, and leaving the chair, he generally, with those of
the members who were so disposed, went into the
Society's library, where tea was prepared.

The Council met in the morning in a separate room
to examine the papers that had been read, or were to be
read, and which were then returned with any remarks
the Council might be disposed to make on them; the
President was supposed to attend, and generally did
attend, the councils, which were mostly held on the days
of the meetings of the Society.

This was the usual process, of the business transacted
by the Royal Society, at the ordinary meetings;
latterly, however, a practice has been introduced by way

of variety, and with a view perhaps to amusement, of
inviting discussion on the papers, that are sent in by individuals to be read at the meetings, an innovation the
policy and the utility of which are questionable. It is
not, I doubt, the best way of encouraging the writing of
gratuitous philosophical dissertations to have them sub-
mitted to a mixed crowd, of whom the majority are probably very imperfectly informed on the subjects treated
of, nor is it likely to increase the number or quality of
contributors.

A similarity of tastes in many points, my having
been also a traveller, and my unfeigned respect for his
character, soon established for me an intimacy with Sir
Joseph Banks, who invited me, with more than common cordiality, to join his Sunday evening conversations
at his house in Soho Square.   This intimacy continued
without interruption to the last days of his life.

About the year 1815, in a friendly conversation one
evening in the library of the Society, Sir Joseph proposed to me to be of the Council, adding that "one of
the Secretaries of the Admiralty should always be in
the Council of the Royal Society."   I concurred in his
view, and was elected in November, 1815, and for many
years I and my colleague, Mr. Croker, were alternately
members of the Council.

An invitation to a dinner or two, with the Royal
Society Club, decided me to have my name put up as
a candidate for a seat in that Club, into which I was
elected in 1808, the first anniversary meeting after my
name was up.   I was well aware that on this occasion
I was patronized by Sir Joseph Banks, but only the

other day was made acquainted with the names of my
supporters, by my good friend Mr. Joseph Smith, the
worthy Treasurer of the Club.

" In reference," he says (in his note to me), " to our
conversation at the Freemasons' Tavern, in regard to
your election into the R. S. Club, I have looked into the
books of the Club; and the names of the members who
were then present, as well as of the candidates who were
then elected, present such a catalogue of illustrious individuals, that I determined to copy the list, thinking
you would feel an interest in seeing *by whom* you were
chosen. I fear we could not present such a splendid
galaxy in these degenerate days."

" At the Anniversary Meeting of the Royal Society
Club on the 14th of July, 1808. Present—

Sir JOSEPH BANKS, President.

Lord Valentia.	Major Rennell.
Mr. Symmons.	Mr. Towneley.
Mr. Barnard.	Mr. Wilkins.
Dr. Simmons.	Mr. Murdoch,
Dr. Maskelyne.	Mr. Horne.
Mr. Cavendish.	Mr. Hatchett.
Dr. Wollaston.	Mr. Best.
Mr. Davy.	Mr. Walker.
Mr. Marsden.	

Mr. M. ROPER, Treasurer.

Mr. Barrow, having been proposed, December 3,
1807, by Mr. Marsden, was elected a member. On the
same day also were elected, Mr. Davies Giddy, Mr.

Benj. Hobhouse, Mr. John Rennie, and Sir James Hall."

It is a melancholy reflection that not one of this " galaxy " now survives, Charles Hatchett, the last that remained on the list, having died on the 10th of March, 1847, when, as next in seniority, I became *Father* of the Club, and in that capacity have drawn up, and now launch, my supplemental tribute to their memory.

In the first place, it may be right to give a short sketch of the formation and plan of the Royal Society CLUB.

It at first consisted, not (as it now does) of the Fellows of the Royal Society exclusively, but separately, under the designation of " The Club of the Royal Philosophers." The following were the original regulations and orders to be observed by the Club :—

" A dinner to be ordered every Thursday for six at one shilling and sixpence a head for eating. As many more as may come to pay one shilling and sixpence per head each ; if fewer than six come, the deficiency to be paid out of the fund subscribed.

" Each subscriber to pay down six shillings, that is for four dinners, to make a fund.

" A pint of wine to be paid for by every one that comes, be the number what it will, and no more unless more wine is brought in than that amounts to."

In 1749 it was resolved " that no stranger be admitted to dine for the future except introduced by the President."

In 1760 it was resolved " that no person shall be chosen a member of this society who shall have three negatives."

From this time the Royal Society Club has regularly progressed, both in numbers and expenses. At present the rule of the club is understood to be, that the number of ordinary members shall not exceed forty, exclusive of the officers, who are the president, the treasurer, the two secretaries, the foreign secretary, and the astronomer royal, all members of the Royal Society, and of the club, without ballot, in virtue of their offices.

The dinner is now ten shillings paid down; and the wine and the tavern bill, exceeding that sum, is settled at the end of the year by the treasurer, who then receives from each member his *quota* of some two or three pounds, according to the amount of the tavern bill.

The rules of election are few:—Every candidate to be proposed by one member and seconded by another.

Candidates to be balloted for, in the order they have been proposed.

None can be elected unless he shall have three-fourths of the votes in his favour.

Every member has the privilege of introducing one visitor.

No visitor shall be admitted on the anniversary of the club.

As there are always more candidates for admission than vacancies, this may in some degree have led to the formation of a new club in 1847, composed of distinguished Fellows of the Royal Society, and others; and to this they have given the name of the Philosophical Club. What their numbers may be, I know not; but the parent club does not appear to have sustained any serious loss by the privation of this learned group.

Had Sir Joseph Banks been still living, this secession
would probably have greatly disturbed him, as the
formation of a new society had done many years ago,
under the designation of the Astronomical Society.
Sir Joseph considered that desertion in a very serious
light, and strenuously opposed it, on the ground that
such an association, by robbing the Royal Society of
many of the first class of its members and affecting to
engross one of its most important departments, struck a
severe blow at its respectability and usefulness. The
new Society was however established, and they selected
and invited the Duke of Somerset to become their first
President. But his Grace, having ascertained the feelings of Sir Joseph Banks on the subject, who appre-
hended from it nothing short of the destruction of the
Royal Society, declined accepting the office: others refused to have their names enrolled in the new society,
because it did not meet with the approbation of Sir
Joseph.

Having occasion one morning to call on Sir Joseph,
I found him very much and very unusually out of
humour. "Have you too," he asked gruffly, "set your
face against me, and joined this Astronomical Society?"
I said "No, Sir; I am not of sufficient calibre, in that
line or any other, to be thought worthy of being invited;
and if I were, I should decidedly decline." "I see
plainly," he continued, "that all these new-fangled asso-
ciations will finally dismantle the Royal Society, and not
leave the old lady a rag to cover her." It was not found,
however, to do much if any harm to the Royal Society,
the same persons in many instances continuing to be
members of both.

I shall now proceed to give brief sketches of the
lives, labours, and characters of the last six Presidents,
and of a few other members of the Royal Society Club,
with whom I had the honour and happiness to be associated, more in social kindness than in any share that
I ever pretended to take in their scientific labours.

Section II.

Sir Joseph Banks, *President of the Royal Society, and of the Royal Society Club.*

The history of this distinguished character is briefly as follows. He was born in Argyle Buildings, London, on the 2nd of February, 1743, O. S., according to a memorandum in his own handwriting—contrary, however, to some other accounts, which represent him to have been born in Lincolnshire, in December of that year, probably because his father, William Banks, had a large property and a fine seat at Revesby Abbey, in that county. There, while a mere boy, he received a domestic education, and in his ninth year was sent to Harrow School, and four years after was removed to Eton, where his good-natured disposition and cheerful temper recommended him to his tutors and made him beloved by his schoolfellows; but the former frequently complained of his extreme aversion from study, especially of Greek and Latin, and of his inordinate love of bodily activity and all kinds of energetic sports. In the course of about twelve months, however, having reached his fourteenth year, his tutor was not a little surprised one day to find him reading at the hours of play; and this change, which had so suddenly been effected in his habits, was subsequently described by himself to his friend Sir Everard Home as having arisen from an accidental circumstance, as he thus related it, and as it was repeated to me by Sir Everard :—

One day, he said, he had been bathing with his fellow Etonians, and, having for a long time been amusing himself in plunging about and trying to swim, on coming out of the water, to dress, he found that his companions had gone away, leaving him alone. Having put on his clothes, he walked slowly along a green lane. It was a fine summer's evening; flowers covered the sides of the path. He felt delighted with the natural beauties smiling around him, and could not forbear exclaiming " How beautiful! would it not be more reasonable, I am sure it would be more agreeable, to make me learn the names and the nature of these plants, than the Greek and Latin I am confined to ?" His next reflection was not quite so agreeable, " that he must do his duty, obey his father's commands, and reconcile himself to the learning of the school."

But this dutiful consideration did not prevent him from immediately applying himself to the study of botany; and, not having any better instructor, he paid some women who were employed in gathering plants— *cullers of simples*, as Sir Joseph himself, in after times, used to call them—and was at that early period quite gratified on receiving such information as they could give him, and for which he paid them as liberally as his finances would admit—sixpence for each article that was collected and brought to him.

Returning home for the holidays, he was inexpressibly delighted to find on his mother's dressing-room table an old torn copy of Gerard's ' Herbal,' having the names and figures of some of the plants with which he had formed an imperfect acquaintance; and he carried it with him back to school—the book, most likely, that the tutor caught him reading. While yet at Eton he continued to enlarge his collection of plants; and he also made a separate one of butterflies and other insects. This, Sir Everard said, he learned from his father.

" I have often," says Lord Brougham in his sketch of Sir Joseph Banks, in his 'Lives of Men of Letters, &c., in the Reign of George III.,' " heard my father say that, being of the same age with Banks, they used to associate much together. Both were fond of walking and of swimming, and both were expert in the latter exercise. Banks always distinguished him; and in his old age he never ceased to show me every kindness in his power, in consequence of this old connexion. My father described him as a fine-looking, strong, and healthy boy, whom no fatigue could subdue, and no peril daunt; and his whole time out of school was given up to hunting after plants and insects, making a *hortus siccus* of the one, and forming a *cabinet* of the other. As often as Banks could induce my father to quit his task in reading or in verse-making he would take him on his long rambles; and I suppose it was from this early taste that we had at Brougham so many butterflies, beetles, and other insects, as well as a cabinet of shells and fossils."

At the age of eighteen Banks lost his father, and thus inherited a substantial patrimony, chiefly of landed property, at an age when the seductions of pleasure are not easily resisted. Fortunately, however, the attractions of science had got possession of his mind, to such a degree as to enable him to resist the temptations of wealth. But he had moreover the blessing of an excellent mother, to advise and encourage him to pursue the track he had chosen in the line of science. The more securely to keep her son in the studious and scientific

line he was so evidently bent on pursuing, she removed
at once, with him and Miss Banks, from Lincolnshire
to the neighbourhood of London, and took up her residence at Chelsea, where she and her son, from their
mutual inquiries, learned that peculiar advantages were
to be procured, especially in the study of botany, in
consequence of the numerous gardens for the cultivation of curious and rare plants of every description.
There was one belonging to the Apothecaries' Company in London—a legacy left by Sir Hans Sloane;
the Botanic Garden; and various nursery-grounds: and
the several curators, superintendents, and others conversant in botany, all held out an irresistible charm for
young Banks.

Having passed a few months among his favourite
plants, it became necessary that he should proceed to
Oxford, to be entered as a gentleman commoner of
Christ Church, as had been decided on shortly before
the death of his father. His love of natural history was
not diminished, and he hoped that means would be
found among the learned of the college to enable him
to pursue it. Botany, however, was the prevailing
object of his wishes; but he found, to his severe disappointment, that no lectures were given by the botanical professor. Though disappointed, he was not
of a temper to be discomfited, much less defeated.
He made application to the botanical professor, and
obtained permission from the learned doctor to find
out and engage a lecturer, the expense to be wholly
defrayed by his pupils. In vain, however, was his
search in Oxford; no one could be found, there, capable
of undertaking the class; but he heard of one at Cambridge, and forthwith went over to that university.

The delight of Banks may be imagined on his falling
in with the very kind of person best suited to his purpose, whom he at once engaged and carried back with
him to Oxford—Mr. Israel Lyons, a learned botanist
and good astronomer. He delivered a course of lectures
on botany, and gave lessons on astronomy to the students, in which Banks of course largely participated;
but Lyons very soon returned to Cambridge, after delivering lectures on both subjects to about sixty pupils:
he was a learned mathematician, and became a cal-
culator for the ' Nautical Almanac.' Banks did not
forget his friend and instructor. He obtained for him
the appointment of astronomer to Captain Phipps, on
his Polar voyage.

Banks continued, with increased zeal, his botanical
studies, which the old Oxonians did not yet much ap-
preciate. He used to tell, in after life, that, when he
entered any of the rooms where discussions on classical
subjects were going briskly on, some one would call
out, " Here comes Banks, but he knows nothing of
Greek:" on which he would quietly say to himself,
" I shall very soon beat you all in a kind of knowledge
that I think infinitely more important." He heard their
jokes with the greatest good humour: these classical
scholars sometimes found themselves at a loss on some
point of natural history, when they would say, " We
must go to Banks."

When Banks came of age (in 1764), and was put in
possession of his valuable estates in Lincolnshire, he of
course left Oxford; but he still continued steadily to
dedicate his time to scientific pursuits, and seemed to
live only for the studies of the naturalist. He went but
little into any other society than that of men known for

their love of science. He was fond of exercise, and that
of hunting and shooting were among his chief recrea-
tions; but so attached was he to angling, that he
would devote days and even nights to that amusement.
It happened that Lord Sandwich had the same taste,
and, their estates being not far from each other, they
became intimately acquainted. It is probable, however, that their first acquaintance commenced when his
Lordship paid a visit to Lady Banks and her son at
Chelsea. There was a story which Lord Brougham
mentions, and which Sir Joseph Banks used to tell, of
a project these two anglers had formed for suddenly
draining the Serpentine, by letting off the water; but
their scheme was discovered the night before it was
to have been put into execution, which had much vexed
these experimental philosophers, who had hoped to have
thrown some new light on the state and habits of the fish.
If such a scheme was really meditated, it was most
probably concocted when Lord Sandwich visited Lady
Banks and her son at Chelsea, and not in Lincolnshire.

In May, 1766, Mr. Banks was elected a Fellow
of the Royal Society, and seems to have considered
it necessary, on having made this first step in scientific society, to attempt something which might show
that he was not unworthy of the honour thus acquired.
He therefore engaged a friend to accompany him in
a small vessel on a voyage to Newfoundland and
Labrador, his main object being to enrich his collection of plants, fishes, and insects, with those of the
northern region. On his way home he called at Lisbon,
and availed himself of the occasion to continue there
his pursuits in botany and natural history.

The voyages which Mr. Banks performed for the promotion and extension of science and philosophical knowledge are known to most readers for whom science and literature, with a knowledge of the globe and of its inhabitants, have any attractions. In taking a brief sketch of these voyages, I shall chiefly refer to those portions of it in which Mr. Banks was individually and deeply concerned.

About the time that Banks had become a Fellow of the Royal Society, there was a spirit of discovery abroad, among the nations of Europe, which was enhanced by the announcement of astronomers of an approaching transit of Venus over the Sun's disc. England was not likely to be deficient in adding her share of zeal, where discovery and scientific research were the objects in question, and she decided on preparing an expedition to the South Seas. On this intention becoming known, Mr. Banks hastened to apply to his friend Lord Sandwich, then at the head of the Admiralty, for leave to join the expedition, with permission to take with him a suite of scientific men and others, which was immediately complied with.

Banks made his preparations on an extensive scale, worthy of his fortune, his reputation, and his zeal for the advancement of science, and particularly that of natural history. He took with him, besides two draughtsmen and four servants, Dr. Solander, the distinguished botanist, a favourite pupil of Linnæus—who had obtained leave from the British Museum, where he was employed as assistant in the department of natural history. The selection of Lieutenant Cook, as commander of the expedition, was most judicious: he was a thoroughbred sailor, a scientific navigator, and a man of great temper and discretion, in every way calculated to establish a spirit of harmony and goodwill among the various classes of persons engaged on the expedition.   Mr. Banks, on his first interview with Cook, expressed himself highly satisfied with his pleasing manners and conversation.

The preparations being completed, the 'Endeavour' sailed from Plymouth on the 25th of August, 1768; and the first land at which they touched, with the exception of a few days at Madeira, was at Rio :—but the viceroy would not permit the gentlemen to reside on shore during their stay here, nor even allow Mr. Banks to go into the country to gather plants, &c.

On their arrival at Tierra del Fuego, Banks, Solander, Monkhouse the surgeon, and Green the astronomer, with their attendants, servants, and two seamen, landed, and set forth to penetrate, as far as practicable, into this cold and rugged country, consisting of " woods before and hills behind," and beyond these of high barren rocks.   Banks, perceiving some of the party not to like the appearance before them, said there could be no doubt of their easily penetrating the wood, and as little of their ascending the hills; and he urged that, as they were now in a new country, which no naturalist had yet visited, they would assuredly be well rewarded for their labour by abundance of alpine plants, and other new objects of natural history.

The party therefore pushed forward, to the high gratification of Banks; but he ere long discovered that it was likely to be at serious risk, from the inclemency of the weather.   It was the month of January, the height of the Antarctic summer; and they soon found that, in this severe climate, their attempts to ascend the mountainous country were likely to be attended with extreme danger,

not only from the excessive cold, but also from the seve-
rity of a snow-storm which overtook them, and in which
three of their attendants actually perished.

Solander, who had more than once crossed the
mountains, that divide Sweden from Norway, was well
aware that extreme cold, when joined with fatigue, pro-
duces torpor and sleepiness that are almost irresistible.
He therefore thought it proper to warn his companions
of the certain fate that awaited them should they yield
to drowsiness. "Whoever," said he, "sits down, will
sleep; whoever sleeps, will wake no more." Yet with
all his knowledge, and the communication of it to the
party, he was himself the first to be so overpowered as
not to be prevailed on to resist lying down; and had not
his more resolute friend Banks come to the spot just in
time to rouse him by main force, he must very soon
have slept the sleep of death: his feet were found
to be so much shrunk that his shoes fell off.

It required no persuasion of any of the party to make
as speedy a retreat as possible from this horrible climate; and Cook, presuming that the state of the weather
would send them back, had made all necessary preparation to get under weigh the moment of their return.
Their next station was Otaheite. Here the first object
was to examine and prepare the astronomical instruments for observing the transit of Venus, which had
been one of the chief motives of the expedition. It was
arranged that Cook and Solander, with Mr. Green,
should be the observers at the fort erected on the island
of Otaheite; that Banks should accompany another
party to be sent, for the like purpose, to the island of
Eimeo.

About six weeks after the conclusion of this important transaction, the 'Endeavour' proceeded on her voyage; cruised for some time among the Society Islands, in which it would be superfluous to state that Banks employed his time in collecting specimens of every branch of natural history which they afforded. During their stay he became, by his kindness and attention, a great favourite among the natives, the good effect of which was soon manifest. The astronomical quadrant was one night carried off from the observatory by some of the islanders. Banks volunteered to go out among them in quest of it—he traced and recovered it. The expedition then proceeded, as directed in Cook's instructions, in search of the Great Southern Continent, a conjectural land, supposed to exist as being a necessary consequence of, and a counterpoise to, the Arctic mountains of the northern hemisphere, a theory of which Dalrymple, sometime Hydrographer of the Admiralty, was the great advocate.

The next land seen by Cook was New Zealand, discovered by Tasman, who named it Staaten Island, but did not land upon it. Cook sailed round it during six months, and diligently explored its coasts. In March, 1770, he commenced his homeward voyage, and proceeded along the east coast of New Holland, then explored for the first time. The natural history of the country, its geography, and its inhabitants, were equally new, and the opportunity of extending researches, to a certain extent, into the interior, was not lost on Banks and his companions. The navigation, as little known as the coast itself, was most perilous; extensive coral reefs, stretching out to a considerable distance from the shore, rising in many places like a wall out of the sea.

In the midst of all the difficulties of an unknown

navigation, and beset with dangers, they still continued to run along the shore, and nearly parallel to it, some twelve or fourteen hundred miles without disaster; when, on the night of the 10th of June, some time after an alarm had been raised, a loud crash but too plainly told them that the vessel had struck. The commander was instantly on deck, and gave his orders with his wonted coolness and precision. The ship had forged over the ledge of the rock, and lay in a hollow within it; the sheathing-boards were seen floating all round her, and were followed by her false keel, so that their fate appeared imminent. Their situation was dismal enough; they were eight or more leagues from the shore; the strictest discipline was preserved, while all in turn were employed in lightening the ship. Two tides had elapsed before she could be got afloat, after everything had been thrown out of her that could be spared—guns, heavy lumber, ballast, stores.

The excessive labour of the crew at the pumps, night and day, so exhausted the men, while the leak still increased, that they were on the point of giving it up in despair, when Mr. J. Monkhouse, one of the midshipmen, suggested their having recourse to an expedient which he had seen practised on a voyage to America, called *fothering*,—that is, drawing under the ship's bottom a sail, in which are stitched down oakum, flax, dung, and other thick and light substances; the motion of the water through the leak draws in the sail, and thus stops the leak. Happily it succeeded, and enabled a single pump to keep the leak under, and to allow of the ship proceeding to a river, to which Cook gave the name of Endeavour. Dr. Hawkesworth, in his account of the voyage, gives high praise to the conduct and the exertions of all, including, in a most particular manner, Mr. Banks and his party.

The joy, however, at so merciful an escape had scarcely subsided, ere that dreadful calamity, the scurvy, began to make its appearance. Among the first attacked was Mr. Green the astronomer, and Tupia, a native of the Society Islands, whom they had taken on board for a voyage to England. And now the value of Mr. Banks became most conspicuous. The country was to be explored for fresh vegetables, for the relief of the sick; and Mr. Banks, with his wonted activity, humanity, and skill, undertook to guide these expeditions. In the course of them he discovered that strange quadruped, since so familiarly known by the name of kangaroo. He found also a supply of fish, turtle, and large cockles, and a variety of vegetables, each of which proved a most seasonable relief.

Nor were the researches of Banks concerning the natives, their manners and habits, less interesting to science. In prosecuting these inquiries, it appears in the commander's journal that his courage was as conspicuous as his activity and his judgment. He would expose himself to collected multitudes of natives when some inadvertent proceeding had aroused their anger, or he would resist them when a thirst of plunder had incited them to threaten; he would visit their habitations unattended by any force whatever; he would sleep for nights together on the ground, at many leagues distant from the crew of the ship, and accompanied only by two or three attendants, regardless of the peril in which he must have been placed, had the natives, living possibly close by, discovered the place of his repose.

On putting to sea with their shattered ship, they experienced a heavy gale, in a navigation beset by reefs of rocks, and shoals, and breakers, which for several weeks threatened them with destruction. Even at this day, the northern part of the eastern coast of New Holland is hateful to navigators; but Cook was now exploring it for the first time, and he laid down in his chart more than two thousand miles of coast, taking formal possession of the country for the British Crown, and giving to it the name of New South Wales.

He thence proceeded to New Guinea, and in October reached Batavia, where the ship underwent a thorough repair. Her bottom, it is stated, was found to be reduced to the thinness of a shoe-sole. Here, in this pestilential climate, the fever broke out among the ship's company, of whom seven died in a few days; and so many were on the sick-list, that not ten men remained fit for duty. Mr. Banks and Dr. Solander were so ill that their lives were considered in danger; Cook also fell seriously ill. When they set sail on Christmas-day, Mr. Banks was carried on board, his life being despaired of. The ravages of the fever continued throughout the voyage; and the "nightly corse" was frequently heard to plunge into the water. Before they reached the Cape, about the middle of March, three-and-twenty had thus perished, including Mr. Green, Mr. Monkhouse, the surgeon, and his brother the midshipman, whose suggestion of *fothering* had saved the ship.

Nothing material occurred in the homeward voyage, and on the 12th of July, 1771, the 'Endeavour' cast anchor in the Downs, delivering up her gallant and prudent commander, with his adventurous companions, to the gratitude and admiration of their countrymen.

The results of this voyage, it is scarcely necessary to observe, were highly important to the sciences of geography and navigation, and to the extension of our knowledge in natural history; and it needs not to be added that Mr. Banks had left nothing unexplored in the whole range of the animal, the mineral, and the vegetable kingdoms of the regions which were visited. But the result of his information was not confined to natural history only; it extended to the manners, the habits, and the condition of the various natives that for the first time had been visited. The record of his observations on all these points, duly and faithfully kept, most essentially helped and improved the compilation of the history of their voyage. Dr. Hawkesworth expressly declared, and with proper feeling, that he was concerned at delivering his account as in the person of the commander of the expedition, acknowledging that, but for the nautical part of the narrative, he should have preferred making Mr. Banks alone the person to speak. This, he says, was in fact proposed to Mr. Banks, "but the proposal was generously overruled."

Important as were the results of this voyage, there still remained much to discover in the South Sea. It had not extended our knowledge of the southern hemisphere beyond the forty-seventh degree of south latitude; and the opinion still prevailed among certain philosophers that a *Terra Australis* must exist far beyond that parallel of latitude, as a counterpoise to the vast extent of mountainous land known to lie beyond the eightieth degree of latitude in the northern part of the globe. To ascertain this point, and to acquire a

greater extent of knowledge regarding the high southern
latitudes, another expedition was set on foot, in the following year, under the command of the same great
navigator, now *Captain* Cook. To him it must have
been highly gratifying as the sure road to fame and
promotion. But it is impossible to reflect, without
astonishment and admiration, on that ardour for the
advancement of science and that noble disregard of
danger, fatigue, and sickness, and of every kind of annoyance and discomfort, which induced Mr. Banks
again, after a few months of repose, to volunteer his
services, though in pursuit of his favourite studies.

To lose no time in making the necessary preparations, Mr. Banks at once engaged Zoffany the painter
and three draughtsmen, hired two secretaries and nine
servants, instructed in the art of preserving plants and
animals. All the books, drawings, and instruments
required for his studies, and all the stores which so
numerous a suite could require, were provided in
profusion; and everything appeared to be in readiness
for his joining the expedition, when he found such
a system was adopted by the Navy Board to thwart
every step of his proceedings, especially on the part of
its chief, the Comptroller of the Navy, Sir Hugh Pal-
liser, whereby his patience was worn out, and his indig-
nation so far excited as to cause him, though reluctantly,
to abandon this enterprise altogether.

Mr. Banks, however, was determined not altogether
to lose the fruits of his extensive and costly preparations, intended for the use of an expedition, which, by
some private intrigue, or hostile individual, he had been
prevented from joining. He prepared for a voyage to
Iceland, hired a small vessel at 100*l*. per month, and

engaged his party, consisting of his tried friends, Dr.
Solander, Dr. Lind, of Edinburgh, and Dr. Von Troil,
a Swedish clergyman of Iceland, besides draughtsmen,
secretaries, seamen, and attendants, amounting in all to
forty persons.

On their voyage out, Banks was desirous of examining that extraordinary basaltic island off the coast of
Scotland known by the name of Staffa. He took some
pains in describing the dimensions and forms of the
columns, making drawings of individual columns and
of the great cavern. This curious paper he conveyed to
Mr. Pennant, who inserted it in his 'Tour to the Highlands of Scotland.' In speaking of Staffa, Mr. Pennant
says, " I wished to make a nearer approach to it, but the
prudence of Mr. Thompson, who was unwilling to venture in these rocky seas, prevented my further search
of this wondrous isle. I could do no more than cause
an accurate view to be taken of its eastern side, and of
those of the other picturesque islands then in sight.
But it is a great consolation to me that I am able to
lay before the public a most accurate account communicated to me through the friendship of Mr. Banks."
And in a note he says, " I cannot but express the obligations I have to this gentleman, for his very kind intention of informing me of this matchless curiosity; for
I am informed that he pursued me in a boat for two miles,
to acquaint me with what he had observed, but, unfortunately for me, we outsailed his liberal intention." He
then proceeds to give Mr. Banks's account of Staffa,
and concludes the fourteen or fifteen quarto pages of
which it consists, full of description and measurements
of the several basaltic columns, with the following
note:—" As this account is copied from Mr. Banks's

Journal, I take the liberty of saying (what by this time that gentleman is well acquainted with) that Staffa is a genuine mass of *basalts* or *Giants' Causeway*,* but in most respects superior to the Irish in grandeur.   I must add that the name is Norwegian; and most properly bestowed on account of its singular structure: Staffa being derived from staf, a staff, prop, or, figuratively, a column." It is stated in a note to Von Troil's Letters, that the fine representations of the basalts of Staffa were printed after accurate drawings executed by Mr. John Frederick Miller, employed by Mr. Banks.

Leaving Staffa, they made the best of their way to their destination, the island of Iceland. It would be superfluous in this place to state with what zeal the stupendous volcanic mountain Hecla, the boiling geyser, and other volcanic streams of emitted water, attracted the attention of Mr. Banks, or the minute observations he registered of those wonderful phenomena which the whole island presented. The copious observations, which we know were made on all these points, he turned over to his friend Von Troil, who afterwards became Archbishop of Upsal, and published a full and interesting account of the island of Iceland, but does not do justice, I think, to the qualities and exertions of Mr. Banks and his companions, whom he dismisses with a too vague and general eulogium.

" You may judge," says he, " how agreeably I spent my time here (Iceland), and with so much more pleasure as our occupations all related to objects entirely new; added to which, I was in society with Mr. Banks and Dr. Solander, the latter of whom is a worthy disciple of

---

* Mr. Pennant might have spared this information, as Mr. Banks had already told him that " Staffa is a Giants' Causeway."

our Linnæus, and unites a lively temper to an excellent
heart, and the former is a young gentleman of an unbounded thirst after knowledge, resolute and indefatigable in all his pursuits, frank, fond of social conversation, and at the same time a friend of the fine arts and
literature: in such company, you will confess, it was
impossible I should have the least reason for regretting
the time spent in this voyage."

Nor had the Icelanders any cause for regret. The
humanity of Mr. Banks was of signal service to these
poor creatures; for when, some years afterwards, they
were in a state of famine, the benevolence and powerful interest of this kind-hearted man brought about
the adoption of measures which absolutely saved the
inhabitants from starvation. We were at war with
Denmark, and had captured the Danish ships, and no
provisions could be received into Iceland. Clausen,
a merchant, was sent to England to implore the grant-
ing of *licences* for ships to enter the island, and through
the active intervention of Sir Joseph, who, as a Privy
Councillor, was an honorary member of the Board
of Trade, the indulgence was granted.

After this expedition to Iceland, Mr. Banks does
not appear to have made any more sea-voyages; but
in 1773, being at Rotterdam, with the Honourable
Charles Greville, they assisted at an assembly of the
Batavian Society, when Mr. Banks communicated to
that society his intention of undertaking a voyage
towards the Arctic Pole, and requested that the principal northern navigators might be desired to put him
in possession of such discoveries and observations as
had been made by their nation, as far as the eighty-
fourth degree of latitude, promising at the same time

to acquaint them with all discoveries that he might be
fortunate enough to make in the course of his voyage.

Twelve years after Mr. Banks had been elected a
Fellow of the Royal Society, and when his fame had
spread throughout Europe and even beyond it, at the
first vacancy that occurred, in November, 1778, he was
chosen President in the room of Sir John Pringle, who
resigned the office at the anniversary, having held it
six years, with great credit to himself and satisfaction
to the Society.

He united the qualifications of physician and philosopher, and wrote numerous treatises and papers in both
capacities. In 1745 he was Physician-general to the
army in Flanders; in 1749 Physician in Ordinary to
the King; in 1752 he wrote observations on the *Dis-
eases of the Army;* on the accession of George III.,
he was appointed Physician to the Queen's household;
in 1763 was elected a Fellow of the College of Physicians in London; and in June, 1766, was advanced
to the dignity of Baronet of Great Britain.

In the six years that he held the chair of the Royal
Society, he appears to have furnished for the Philosophical Transactions two papers each year. He had
a splendid mansion in Pall Mall, where, in addition
to a general hospitality to the literary world, he on
Sunday evenings held *Soirées* or *Conversations,* which
were attended by all the learned of Great Britain and
of foreign nations who happened to be in London. But
his reign was not undisturbed. It is well known that
disputes on controversial points frequently occur in
learned societies, and that they are generally carried
on with more acrimony than among ordinary people.

A great contention had arisen in the Royal Society, which of course involved the President and annoyed him grievously: it concerned the question whether electrical conductors placed on buildings are the most efficacious in preserving them from the destructive effects of lightning when made with points or with knobs; a dispute which, simple as it appears, was carried on with great vehemence by the two contending parties —the one, Mr. Nairne, supporting Dr. Franklin's system of *points*; the other, Wilson, contending for the *knobs*. George III. is said, I know not on what authority, to have earnestly entreated Pringle to support Wilson, and that Pringle's answer to the King was, "Sir, I cannot reverse the laws of nature." This seems to me very questionable: the good King may, for aught I know, have formed a personal opinion in favour of *knobs*, but he was much too sensible a man to think that royal solicitation could decide a point of practical and experimental science. There can be no doubt of, and indeed Wilson's own experiments seem to establish, the superiority of the *point* principle adopted by Sir John Pringle: but the dispute, on it, was supposed by many (erroneously, I think) to have had the effect of inducing the resignation of the President; for his resolution of resigning had previously been made to his friends, on the ground that his advanced age and that the impaired state of his health required it. In 1781 he left London to make an excursion to Scotland, found himself unwell towards the latter end of the year, and died on the 18th of January, 1782, greatly respected as an able and honest man.

Mr. Banks followed the example of his predecessor in

many respects. On his return to England he took up his abode at a mansion in the south-west angle of Soho Square, where his extensive library, embracing every subject of literature, and containing volumes embellished and illustrated with prints, maps, and charts, besides an infinite variety of drawings of subjects in natural history, more especially in botany, was accessible to the student and amateur. Every Thursday morning a breakfast was prepared for all who would come to partake of it, and this afforded the visitor an opportunity of consulting him on any favourite study he might be engaged upon, or on the best works in which he might expect to find the information of which he was in quest. Foreigners too were his ever welcome guests, and it was his delight to be surrounded by the cultivators and the promoters of science of all countries, and in all its branches. Each Sunday evening after dinner he held a *conversation*, like Sir John Pringle, at which the *literati* of all nations were to be met; curiosities of every description were brought by the visitors and exhibited; and each new subject, book, drawing, animal, plant, or mineral, each invention of art or science, was sure to find its way to the house of Mr. Banks in Soho Square.

The general reputation in which Mr. Banks was held, the fame of his voyages, his wealth, and his liberal and courteous reception of all, and more especially of foreign visitors, gave him an extensive social influence, and made his house an exceedingly agreeable centre of the literary world. He felt the importance of his position, and devoted himself with his accustomed ardour to the various duties of so peculiar a station, and for some years his administration of the duties of President of the Royal Society met with a very general approval. But it has almost always been found that he who honestly undertakes to reform the abuses of a public department must lay his account to the creation of enemies, and so it happened in the case of Mr. Banks.

He had not long been in the chair of the Royal Society before he perceived, and was at once determined to put an end to, certain manifest abuses. One of these was the indiscriminate admission of candidates, to be balloted for election into the Society, who had no pretensions to become members of that body. On one occasion, after a somewhat rigid examination, some half-dozen unworthy candidates were black-balled, one of whom was stated to be the patentee of a new water-closet. Mr. Banks therefore thought it right to announce to the Council, the secretaries, and members of the Society, his determination to watch over the applications for admission to the balloting list. The better to secure that none but proper persons should be admitted, he said that " previous to the election he should speak to the members who usually attended, and give his opinion freely on the merits of the candidates, and that, when he considered a rejection proper, he should not hesitate to advise it." Many attacks, as may be supposed, were made on him in consequence of this regulation, and of the number of rejections that followed.

But an occurrence of a different kind was the occasion of vehement dissensions, that disturbed for a time the peace of the Society. The office of Foreign Secretary had been conferred on Dr. Charles Hutton, a mathematician of distinguished reputation, whose official duties, as Professor at the Royal Academy of Woolwich, obliged him to reside there. This gave rise to a complaint of a neglect of his duties as Secretary. The Council of the Society, under the influence, it was said, of the President, passed a resolution recommending that the Foreign Secretary should reside in London; and by this measure obliged Dr. Hutton to tender his resignation. This occurrence caused a great sensation in the Society. The discontent with the President's administration, on account of the circumstances we have mentioned, which had for some time been smouldering, now broke out in a flame. Dr. Maskelyne, the Astronomer Royal, considered his friend Dr. Hutton as ill treated, and so did others; the most active and noted of whom was Dr., afterwards Bishop, Horsley.

Dr. Horsley, finding himself supported by Dr. Maskelyne and Baron Mazeres, hoisted a standard for the mathematical sciences, in opposition to natural history, which the President and his especial friends chiefly cultivated. A party was got up, who raised, what I cannot but think, an absurd and an unfounded clamour that the mathematics were neglected, and that botany alone was patronized; hoping to eject Mr. Banks, and to place, it was said, Horsley in the chair. This latter object was denied; but, however that may have been, various motions were made which would have had the effect of displacing the President, had they not all signally failed.

The excellent qualities of the President whom this victory kept in the chair were more strongly exhibited by the temper with which he regarded the opposition. The sketch of his character given by Lord Brougham is true to the life. "He showed no jealousy of any rival, no prejudice in any one's favour rather than another's. He was equally accessible to all for counsel and for help. His house, his library, his whole valuable collections were at all times open to men of science, while his credit both with our own and foreign governments, and, if need were, the resource of his purse, was ever ready to help the prosecution of their inquiries."

" Many circumstances concurred to give to Mr. Banks the power which he so largely exercised of patronizing and promoting the labours of scientific men : his ample fortune—the station which he filled in society—the favour which he enjoyed at Court and with the Ministers of the Crown—the fame of his voyages—his indefatigable industry—his ever-wakeful attention to the representations and requests of the student—his entire freedom from all the meaner feelings which mere literary men are but too apt to entertain one towards another—his great natural quickness and unerring sagacity, never leaving him long to seek for the point of any argument, nor ever suffering him to be diverted by plausible errors or designing parties—his large and accurate knowledge of mankind, and of men as well as of man—the practical wisdom which he had gathered from extensive and varied experience—all formed in him an assemblage of qualities, natural and acquired, extrinsic or accidental, and intrinsic or native, so rare as had hardly ever met together in any other individual."

His own studies continued to be, as they had always been, devoted to natural history ; and botany was still the portion of it which he chiefly loved to cultivate. During the greater part of his life, his time and his fortune were assiduously employed on the preparation of a magnificent series of botanical drawings and engravings. But

he never retained any of these, locked up as it were, for
his own gratification. He kept Mr. Baur, at Kew, constantly occupied in making copies of Australian and
other South Sea Island plants. The whole life of this
excellent draughtsman was employed in drawing and
colouring the most exquisite flowers, for which Sir Joseph
kept him in liberal pay; nor did he forget poor Baur
in his will. He gave little or nothing under his own
name to the world in science or literature, though not
deficient in either. Indeed we are told his habitual
indifference to literary fame made him so slow to publish, that he is believed to have constantly given over
to other cultivators of the same studies the fruits of
his own labour, as those fruits were ripened and
gathered in; and while all men's books were crowded
with his designs, and all men's inquiries promoted by
the stores of his knowledge, he alone reaped no fame
from his researches, nor profited by the treasures which
he had amassed.

Dr. Horsley's qualifications for the Presidency were
very different: the haughtiness of his manner had become repulsive, even to his own party. Dr. Kippis,
in alluding to him, says, " The manner which he
assumed during the late dissensions will not easily be
forgotten. The impression will long remain in the
minds of the members, of the power of voice, and
the energy of words, with which his denunciations
were delivered. The high tone he adopted went beyond
the usual custom of public debates." More than once
it appears in his speeches he called the Committees
packed juries; and it is well known that, when he perceived a defeat approaching, he threatened the secession of the mathematical party, and exclaimed, "The

President will then be left with his train of feeble amateurs and that toy* upon the table—the ghost of the Society in which philosophy once reigned, and Newton presided as her minister." Yet Newton, the philosopher, astronomer, and mathematician, had been preceded in the chair by Lord Somers, whose eminence was certainly not derived from science or philosophy, and was succeeded by Sir Hans Sloane, whose acquirements had no sort of connexion with those of Newton; nor were the supporters of his election of that class.

In truth, the chair of the Royal Society does not require to be perpetually and exclusively filled by men of science, or by persons elevated in any one particular department of science. The President should be conversant in general knowledge, especially in the knowledge of the world, courteous and agreeable in his manners and conversation, ready to oblige and to forward to the best of his power the objects brought to the consideration of the Society; in short, to follow the example of Sir Joseph Banks, in promoting intercourse among the members at certain fixed times set apart for that purpose; and, above all, should select men of science and literature for his Council, on the character and conduct of whom the progressive success of the Society must mainly depend. A President devoted to *one* science is not the fittest for the Royal Society, and it was no doubt with this conviction that the attempt was made to fix upon Sir Joseph Banks the character of a botanist rather than that of a patron of science in general.

* Charles II., when he incorporated the Society, lent them a mace from the Tower, as is done to the Houses of Parliament. Hence Bishop Horsley's imitation of Cromwell's contemptuous designation of the mace of the House of Commons as a bauble.

We have seen that Sir Joseph Banks bore his victory without elation, and displayed no jealousy or prejudice. Indeed he very soon showed his magnanimity on this trying occasion. At the following anniversary he thus addressed the meeting:—

" From the appearance of our present meeting I will venture to foretell that our disputes are at an end— that the gentlemen from whom I have had the misfor- tune to differ in opinion will abide by the decisions of the Society, which they have repeatedly taken, and will agree with me in the determination to throw a veil of oblivion over all past animosities, uniting once more in sincere efforts towards the advancement of the Society, the honour and reputation of which we have all equally pledged ourselves to support."

" But," he adds, " enough of dissension ; a word, never more, I sincerely hope, to be heard within these walls, dedicated as they are, by a generous monarch, to the service of science. Peace and harmony should ever be found within them; for under their influence alone can science flourish among those who profess to cultivate it.

" Let us unite once more, my friends, to fulfil the wise purposes of our liberal patron and benefactor, and resume at the same time the prudent conduct of our predecessors, who for more than a century supported the honour of this Society unsullied, and have be- queathed it to us as pure as they received it. They never failed to sacrifice such differences as rose among them to the good of the general cause, in which they felt themselves equally embarked ; for although some individuals among them have heretofore indulged their feelings by appealing to the public, when they imagined the welfare of the body at large was in danger, they never once attempted to convert the meetings, instituted for the advancement of knowledge, into assemblies of debate and controversy."

Sir Joseph Banks was in fact so well supported by all the eminent men of the day in the arts and the severer sciences, that he was anxious to have it put to the test how he stood with the Society at large after Dr. Horsley's opposition, and that the ballot should be appealed to for that purpose. The ballot was accordingly taken by the members who came down on the occasion, and a vote of confidence, "approving of Sir Joseph Banks as their President, and of supporting him in his office," was carried by the large majority of 119 to 42. Horsley, after this defeat, withdrew his name from the Society.

That a man of Banks's standing in society—a man of wealth and influence as a country gentleman, who had employed a great portion of that wealth for the benefit of science and for his country's renown, in exploring foreign regions of the world, at the risk of health and the sacrifice of every comfort and convenience,—that such a man, on his return, should instantly have received all the honours that a grateful country could bestow was reasonably to be expected; and they certainly were ultimately, though gradually, bestowed. He had been elected a member of the Royal Society by ballot in 1766, as any decent tradesman might then have been. After twelve years of attendance, and the many benefits bestowed on the Society, in 1778 he was chosen President. In 1781 he was created a Baronet; in 1795 he was invested with the Order of the Bath; and in 1797 he was advanced to the dignity of a Privy Councillor. In 1802 he was chosen a foreign Member of the Institute of France.

In the *Memoirs of the Life of Sir Humphry Davy*, by his brother, are given a number of "Characters," by Sir Humphry, among which that of Sir Joseph Banks occurs.

"He was a good-humoured and liberal man, free, and various in conversation, a tolerable botanist, and generally acquainted with natural history. He had not much reading and no profound information. He was always ready to promote the objects of men of science; but he required to be regarded as patron, and readily sanctioned gross flattery. When he gave anecdotes of his voyages, he was very entertaining and unaffected. A courtier in character, he was a warm friend to a good king. In his relations to the Royal Society he was too personal, and made his house a circle too like a court."

I think Sir Humphry has not done justice to Banks in the character which he has thus drawn: "a *tolerable botanist*, a lover of *gross flattery*, a house *like a court*," are expressions, in my opinion, unfounded and unjust. Sir Joseph was in many respects an extraordinary man. Among these was his ardent desire to arrive at philosophical truths, even at the expense of personal inconvenience. A trifling instance of this occurs to me. Dr. Sir Charles Blagdon had instituted a course of experiments to determine the power of human beings to exist in rooms heated to an excessive temperature. Sir Joseph Banks was one of the first who plunged into a chamber heated to the temperature of 260° of Fahrenheit, and was taken out nearly exhausted.*

Sir Joseph Banks had become a great martyr to the gout, which grew to such an intensity as to deprive

* Sir Francis Chantrey entered a furnace for drying moulds, and remained in it two minutes, at the temperature of 320°.

him entirely of the power of walking, and for fourteen or fifteen years previous to his death, he lost the
use of his lower limbs so completely as to oblige him
to be carried, or wheeled, as the case might require, by
his servants in a chair : in this way he was conveyed to
the more dignified chair of the Royal Society, and also
to the Club—the former of which he very rarely omitted
to attend, and not often the latter ; he sat apparently so
much at his ease, both at the Society and in the Club,
and conducted the business of the meetings with so
much spirit and dignity, that a stranger would not have
supposed that he was often suffering at the time, nor
even have observed an infirmity, which never disturbed
his uniform cheerfulness.

It happened but very rarely that he was compelled to
forego the pleasure of personally appearing at those
Sunday evening assemblies in Soho Square, which, to his
numerous guests, including, as I have already said, many
foreigners, were really a great and a rational treat.
Every new discovery in the range of natural history,
every new and ingenious specimen of art, every curious
and useful invention, was sure, at some one of these
meetings, to be exhibited at Sir Joseph Banks's house,
which was itself a repository of art, science, and literature.   On one of these evenings a discussion took place
among the physiologists concerning the noxious or nu-
tritive quality of the flesh of certain animals, taken as
food by human beings ; and, as usual, various opinions
and speculations were hazarded ; one person observed that
the flesh of a shark, which delighted in human food,
must be at least unwholesome.   Sir Joseph Banks
said, " I have eaten shark, and I believe I have eaten
of every fish that inhabits the ocean, at least of all

that we know of, out of pure curiosity; and moreover have tasted the flesh of most animals that inhabit
the earth; and am not aware of having suffered any
harm or inconvenience from any of them." I happened to ask Sir Joseph if he had ever tasted the flesh
of a hippopotamus. "I augur," he said, "from that
question, that *you* have." "Yes, I have; and if you,
Sir Joseph, should not have eaten the flesh of that huge
animal, I can now assure you it is as good as, and very
like, a pork-chop." Sir Joseph immediately observed,
".I have no doubt you are right, for the *hippopotamus*,
though we so call him, is certainly more like a hog than
a *horse*. You have the whip-hand of me; for you
have twice doubled the Cape: I have only touched at
it on our voyage home."

As the gout increased his difficulty of locomotion,
Sir Joseph found it convenient to have some spot to
retire to in the neighbourhood of London, and fixed
upon a small villa near Hounslow Heath, called Spring
Grove; consisting of some woods and a good garden laid
out with ornamental shrubs and flower-beds, and neatly
kept, under the inspection of Lady and Miss Banks.

I had the pleasure of receiving a general invitation
to Spring Grove, and was always glad to avail myself
of the opportunity whenever I could manage to get
there. The ladies were most agreeable companions,
and Sir Joseph was cheerful and much at his ease,
which was not always the case when in town. He
still, however, regularly attended the meetings of the
Society when not under the visitation of so severe a fit
of the gout as to render him helpless.

He one day told me that he had some chestnut-trees,
but he supposed the fruit, like all our English chestnuts,

would be found good for nothing. I said—Mr. Murdoch, who had passed a great part of his life in Madeira, mentioned to me that, till they grafted the tree in that island, the fruit was as bad as ours; but that, when the practice of grafting became common, the Madeira chestnuts were as good as those of Spain. " I shall immediately send for grafts to Spain," was Sir Joseph's answer.

The ladies had a small pond in the garden well stocked with gold-fish, and one day Lady Banks said, " I must tell you a droll story about these fish. Sir Joseph had a visit from two young Americans, who brought letters of recommendation. They were shown round the grounds, and, coming to the edge of the pond, I asked if they had any fish of the kind in their country. One of them said, ' No, we have not.' The other said, ' They appear to be a species of herring; but I never heard of, and never till now saw, a red herring alive!' Now," said Lady Banks, "you must not let Sir Joseph know that I have told you this story, for he says nobody will believe it; but, believe it or not, I care very little—it is true."*

In this latter part of his life I had frequent access to him, and more especially from the time that the preparations for the Arctic voyages commenced. Sir Joseph took a great interest in them. I had published a small 'Chronological History of Arctic Voyages,' which caught his attention strongly, as reminding him of the adventurous spirit of our brave old navigators, whose voyages were his choicest books in early days. " You must come," he said, "and dine with us one day, to meet old

---

* If so, it was evidently meant as a pleasantry.

Scoresby, the celebrated Greenland whale-fisher, who has given me more information about the ice in those regions than any other I have conversed with." I met him, and afterwards received much information from him.

It has been a subject of deep regret, and it is almost unaccountable, that we should possess no regular and authentic Life, and scarcely even any short and imperfect history, until Lord Brougham's appeared, of so eminently distinguished, so actively useful, and, in every respect, so extraordinary a man, whose character and energetic devotion to the pursuit and dissemination of science, is so well known; and it is to be regretted that the renown of such a man should not have found a countryman of his own to be his biographer, that task having been left to a French *savant*, Baron Cuvier, who has enrolled his name in the *Eloges Historiques* of distinguished men. Lord Brougham remarks in his account of Sir Joseph Banks, that, " Although his active exertions for upwards of half a century left traces most deeply marked in the history of the natural sciences, and though his whole life was given up to their pursuit, it so happened that, with the exception of one or two tracts upon agricultural and horticultural questions, he never gave any work of his own to the world, nor left behind him any manuscripts beyond his extensive correspondence with other cultivators of science. It is from this circumstance that not even an attempt has ever, as yet, been made to write the history of Sir Joseph Banks."

His Lordship is, I believe, not quite correct in saying that Sir Joseph Banks did not "leave behind him any manuscripts beyond his extensive correspondence;" for, calling on him one day, I found him in his small library, and after some short conversation, he said, "I am very helpless, as you know; but I wish to show you in what state I have had my papers arranged, and ready for any purpose my executors may decide upon. Take this key, which opens that closet on the left of the fireplace, and you will see how they are assorted in bundles, and labelled; it has been a tedious job, but I had good assistance." I said I was delighted to see his valuable papers in such a state of preparation, and hoped ere long that he and the world would be gratified by possessing their contents in another shape. He guessed my meaning, and said emphatically, " Certainly not in my lifetime."

I happened to mention this interview, not many months ago, to Mr. Robert Brown, with a remark that I took for granted it was to him (Brown) that Sir Joseph had referred. His reply was, " Positively not; Sir Joseph himself arranged the papers just as you saw them;" and I suppose they remained as they were till his death, and that they were the whole or a part of those which were put by the executors into the hands of a gentleman, as materials for a Life of Sir Joseph, and with a view to publishing such as might be of interest, but after a lapse of eighteen or twenty years nothing had been done, and the papers were returned to Sir E. Knatchbull, one of the executors.

It is much to be wished that Mr. Brown may be prevailed on to undertake the work, too long deferred. He may be said to have succeeded Mr. Dryander,*

* Dryander was a learned Swede whom Sir Joseph attached to himself in the character of secretary and librarian. He resided in Sir Joseph's house, and did the honours of his conservatories and library. He died there in October, 1810, aged sixty-two.

whose death occurred ten years before that of Sir
Joseph, on which occasion, I received from the latter a melancholy letter, in which he says, " Poor
Dryander is no more; I have lost in him my right arm."
If Mr. Brown should consent to edit Sir Joseph's papers,
he will not lack any assistance he may require from me,
so far as I may be able to give it.

There is no doubt that this mass of papers contains,
in addition to his correspondence, several original
writings of his own.   One of them, I am assured from
the best authority, is curious, interesting, and well
written, being a dissertation on the history and art of
the manufacture of porcelain by the Chinese, illustrated
by a very select and extensive collection of choice and
variegated specimens, that were in the possession of
Lady Banks.   It has been said, his executors were
desired not to make public certain of his papers, and
to destroy others; that might, no doubt, be a very
proper injunction ; but it would be no reason (quite the
contrary, indeed) for withholding the other papers not
included in that restriction.   I have already alluded to
what he gave to Pennant, to Von Troil, and to Hawkesworth; but I shall add a few words on this topic, with
reference to the probable value of the papers he may
have left.

Hawkesworth, in his Introduction to Cook's Voyage,
tells us that—

" Mr. Banks kept an accurate and circumstantial
account in his journal of the voyage, and was so obliging
as to put it into my hands, with permission to take out
of it whatever I thought would improve or embellish
the narrative.

" In the papers so communicated, I found a great

variety of incidents which had not come under the notice
of Captain Cook, with descriptions of countries and
people, their productions, manners, customs, religion,
policy, and language, much more full and particular than
were expected from a gentleman whose station and office
naturally turned his principal attention to other objects.
For these particulars, therefore, besides many practical
observations, the public is indebted to Mr. Banks; also
for the designs of the engravings which illustrate and
adorn the account of this voyage, being copied from his
valuable drawings, and some of them from such as were
made for the use of the artists at his expense."

And he concludes,—

" It is indeed fortunate for mankind when wealth
and science, and a strong inclination to exert the
powers of both for purposes of public benefit, unite in
the same person: and I cannot but congratulate my
country upon the prospect of further pleasure and advantage from the same gentleman to whom we are
indebted for so considerable a part of this narrative."

Notwithstanding Sir Joseph's liberal feelings, and
that he was always ready to give any assistance in his
power to others; yet, as I have remarked, he had an
inveterate dislike to make any display of his literary
acquirements, variable and extensive as they were, by
publishing them under his own name. It was but the
other day that, on dipping into Sir William Hooker's
'Journal of a Tour in Iceland in the Summer of 1809,'
I found that he quoted passages from the "*Private MS.
Journal of Sir Joseph Banks.*"

I asked Sir William where he had been fortunate
enough to obtain these journals. The following is a
copy of his reply:—

" West Park, Kew, 18th October, 1848.

"My dear Sir John,

"The MSS. from which I quoted in my 'Tour in Iceland' (and to which you allude) were from Sir Joseph's private journal, which he was so good as to lend me for the purpose, and which were in his own writing, and always kept, I feel sure, among his own private papers. He observed to me that, though Dr. Von Troil had the use of those MSS. for his account of the voyage, yet that a great deal of interesting matter was not made use of by the Doctor: and so I found it, and then made the extracts, and of course returned the papers. There was also a fine set of drawings in a large portfolio, and these belonged to the journal; and it is possible they may both have gone to the British Museum.

" Believe me, &c.

" W. J. Hooker."

I shall now extract a passage from Sir Joseph Banks's journal, in which he describes his ascent of Mount Hecla, and I do so because I have never before found it in print; and indeed I had been told that he never was near this mountain, much less on its summit.

"We ascended Mount Hecla," says Sir Joseph, "with the wind blowing against us so violently that we could with difficulty proceed. The frost too was lying upon the ground, and the cold extremely severe. We ourselves were covered with ice in such a manner that our clothes resembled buckram. On reaching the summit of the first peak, we here and there remarked places where the snow had been melted, and a little heat was rising from them, and it was by one of these

that we rested to observe the barometer, which was
24·838, thermometer 27. The water we had with us
was all frozen. Dr. Lind filled his wind-machine with
warm water: it rose to 1·6, and then froze into specula,
so that we could not make observations any longer.
We thought we had arrived at the highest peak, but
soon saw one above us, towards which we hastened.
Dr. Solander remained with an Icelander in the intermediate valley; the rest of us continued our route to
the summit of the peak, which we found intensely cold;
but on the highest point was a spot of three yards in
breadth, whence there proceeded so much heat and
steam that we could not bear to sit down upon it:
hydrometer 9·25, barometer 24·722, thermometer 38.
The last eruption of 1766 broke out on a sudden,
attended by an earthquake. A south wind carried a
quantity of ashes to Holum, a distance of a hundred
and eighty miles! Horses were so alarmed as to run
about till they dropped down through fatigue, and the
people who lived near the mountain lost their cattle,
which were either choked with ashes or starved before
they could be removed to grass. Some lingered for a
year, and on being opened, their stomachs were found
to be full of ashes."—*Sir Joseph Banks's MS. Journal.*

I think I have said enough to invalidate the opinion of
Sir Joseph Banks leaving no written documents behind
him, the products of his own pen, which some persons
have so absurdly exaggerated as to say that he abstained
from committing himself to paper, on any subject connected with science or literature, from a consciousness
of his deficiency in such studies. That he entertained
a too modest sense of his own acquirements is very
true, but he never carried it to the extent of declining

E

to give his opinion on any topic that was presented to him, and those opinions were, as far as I have ever seen them, highly valuable. I very lately accidentally fell in with a printed letter of Sir Joseph, written in the decline of life, which so well explains my view of his character on this point, that I think proper to insert it here, to show that not only could he write, but write well.

Sir Joseph Banks to Sir James E. Smith.

" My dear Sir James,

" My chief reason for troubling you with this is to tell you that I have paid obedience to your mandate, by reading your article on Botany in the Scotch *Encyclopædia*, which, conceiving it to be an elementary performance, I had neglected till now to peruse.

" I was highly gratified by the distinguished situation in which you have placed me, more so, I fear, than I ought to have been. We are all too fond of hearing ourselves well spoken of by persons whom we hold in high regard ; but, my dear Sir James, do not you think it probable that the reader who takes the book in hand, for the purpose of seeking botanical knowledge, will skip all that is said of me, as not at all tending to enlarge his ideas on the subject ?

" I admire your defence of Linnæus' Natural Classes. It is ingenious and entertaining, and it evinces a deep skill in the mysteries of classification, which must, I fear, continue to wear a mysterious shape, till a larger portion of the vegetables of the whole earth shall have been discovered and described.

" I fear you will differ from me in opinion, when I fancy Jussieu's Natural Orders to be superior to those of Linnæus. I do not however mean to allege that he has even an equal degree of merit in having compiled them. He has taken all that Linnæus had done as his own; and having thus possessed himself of an elegant and substantial fabric, has done much towards increasing its beauty, but far less towards any improvement in its stability.

" How immense has been the improvement of botany since I attached myself to the study; and what immense facilities are now offered to students, that had not an existence till lately! Your descriptions, and Sowerby's drawings of British plants, would have saved me years of labour, had they then existed. I well remember the publication of Hudson (in 1762), which was the first effort at well-directed science, and the eagerness with which I adopted its use.

" I am, &c.,

" Jos. Banks."

Not long before Sir Joseph's demise he sent to me to say it would be really a kind act if I could contrive to spend a whole day with him at Spring Grove. I did so; but, as I anticipated, found him much altered in his looks, and more than usually languid; yet he soon grew cheerful, and laughed and joked with Lady Banks, and with one or two others who composed the small party. He talked much of the Royal Society, of the pleasures and annoyances he had experienced in the chair for more than forty years; as also of the great kindness shown in retaining him in his old age, when no longer of any use; and he regretted exceedingly to find that Dr.

Hyde Wollaston would not consent to be put in nomination for the chair,—" so excellent a man, of such superior talents, and every way fitted for the situation!" " Mr. Humphry Davy is a lively and talented man, and a thorough chemist; but, if I might venture to give an opinion, which I do not hesitate to give to you, he is rather too lively to fill the chair of the Royal Society with that degree of gravity which it is most becoming to assume."

He gave his opinion briefly of some other Fellows; spoke in praise of the Royal Society Club, which he considered as a great relief to the monotony of the Society's meetings; and mentioned the names of several whose conversation, alternately serious or lively, combined so much information and good humour. Lady Banks told me, on going away, that my visit had roused Sir Joseph very much, and trusted I would renew it; but his rapidly increasing illness prevented me from having the satisfaction of ever seeing him again. From the 16th of March, 1820, to the 19th of June of that year, his strength and spirits had been gradually giving way, and he died, at the advanced age of seventy-seven.

I felt most deeply the loss of Sir Joseph Banks. For twelve years I had experienced nothing but kindness from him: whether at the Society or the Club, at his house in Soho Square, or at his favourite retreat at Spring Grove, I was always a welcome guest. During the last three years of his life, when the first Arctic expeditions were in discussion and in progress, and in which Sir Joseph took a deep interest, I was very much with him. One day I found him highly elated, and he told me, laughing, that had I come a little sooner I should have enjoyed his interview with Mr.

Lewis Way, the converter of the Jews " I said to
him, ' Way, you do not pretend that by going to Jerusalem you will succeed in making the Jews Christians?'
' Yes,' he said, ' I do; and I will tell you what, Sir
Joseph, I shall go to Jerusalem, and carry my point,
long before you will ever reach the North Pole.' "

Section III.

William Hyde Wollaston, M.D., *Acting President.*

William Hyde Wollaston was one of the most remark-able men of his day, for his extensive acquirements in almost every branch of knowledge, at least in every de-partment of art and science, as well as for being deeply versed in natural philosophy. Of the seventeen sons of the Rev. Francis Wollaston, rector of Christ Church, he was the second, and was born the 6th of August, 1766. Where he was brought up, and by whom, does not ap-pear; but among such a number of sons, it is most pro-bable it was at home, and under the immediate eye of his father. His education, however, was completed at Caius College, Cambridge, where he took his degree of M.D. in 1793.

I am not aware of any Memoir that gives an account of his early life, but the age at which he took his degree would not seem to indicate any precocity of knowledge in the course of the first seven-and-twenty years; nor does it appear that his ambition yet extended beyond the practice of physic, in which he attempted to establish himself in the small town of Bury St. Ed-munds. Not meeting there with much success, he speedily removed from thence and resolved to try his fortune in the great theatre of London, where all the young aspirants for fame expect, though but a few obtain, the contemplated prize. The situation of physician to St. George's Hospital becoming vacant,

Dr. Wollaston offered himself among a number of candidates; but the election fell on Dr. Pemberton, which so mortified Wollaston that he determined at once to "throw physic to the dogs;" with a resolution to abandon the profession altogether, and never to write another prescription, "not even for my father." This resolution would appear more hasty than the occasion required, and not quite characteristic of the man, who is said, in after life, never to have abandoned any plan or project which he had once decided upon accomplishing. Nor should he have felt any serious mortification; he had then acquired little or no experience in the medical profession, whereas Pemberton had already established a character; and St. George's Hospital was of a magnitude to demand an experienced and first-rate physician.

Wollaston, though defeated, was by no means discomfited; his resolute disposition bore him up, and decided him at once to set about a regular course of study, but in a different line, and to direct his attention almost exclusively to the pursuit of natural philosophy in all its various departments, but more especially to the practical operations of chemistry and to mineralogy. Being an ingenious mechanic, he shortly succeeded in making his own chemical and other instruments with simplicity, and at the same time with admitted improvements; for example, he contrived a small galvanic battery, so minute, yet so perfect, as to be contained within a tailor's thimble, respecting which I shall presently give an interesting anecdote. Among his many neat and useful inventions was that of the camera lucida, the merits of which are well understood by the teachers and pupils of the art of drawing.

Among the many discoveries, which he had the good

fortune to make, was that of rendering platinum (a
metal till then but imperfectly known, and unmanage-
able) not only malleable, but easily convertible into
various shapes and uses never before thought of, and
capable of being drawn out into a wire as fine as a hair,
—so delicately fine as to be imperceptible nearly to
the naked eye.   By his expert management of this
metal, and its application not only to the various instruments of his own invention, but to many of the
common purposes of life, the malleable platinum was
supposed to have produced him, from first to last, a
handsome fortune; not less, it has been said, than thirty
thousand pounds. It did more. This invention, together
with his numerous other inventions and discoveries,
most of which were practically useful, and therefore
profitable, raised Wollaston's reputation and consequence
to the highest pitch.   In 1804 he made known the two
new metals, palladium and rhodium, which he discovered
to be produced from the ore of platinum.   He also
proved that tantalum was identical with columbium,
previously discovered by Mr. Charles Hatchett.

Wollaston had for some time been elected a Fellow
of the Royal Society, and in November, 1806, became
second secretary of the Council. His communications
to the Philosophical Transactions were numerous and
important; and at the anniversary of the 30th of November, 1828, the Royal Society awarded to him the
Copley Gold Medal, for his essay ' On rendering Platinum malleable.'   Eight years before this, on the
illness and death of Sir Joseph Banks, the Council of
the Royal Society proposed to nominate Wollaston for
the succession to the chair, but he declined being put
in nomination, in so decided a manner as to make it

expedient to look out for some other candidate. He said, however, that until they could pitch upon a proper person he had no objection to accept it, as Acting President, for a few months, until the regulated time arrived for the election of a President.

Among the various investigations of Dr. Wollaston was a minute inquiry into the nature of crystallization generally, and especially into the constitution of the diamond and other precious stones. There was at this time a story circulated by certain persons, not the most friendly disposed to Wollaston, about a diamond which he happened one day to see, in passing by one of those second-hand shops where all kinds of curiosities are exposed for sale. His attention being caught by a brilliant pebble, he went into the shop to look at it; he discovered it to be a diamond with a large flaw, and, after a close inspection, inquired of the shopman what price he had put on that damaged article. He named the price, which Wollaston paid and took away the pebble, called on his way at a lapidary's of his acquaintance, and said he wished him to make a cut in the stone, just as he should direct him. When told what to do, " That," said the man, " would utterly destroy the value of the stone;" but Wollaston said, " Do as you are directed." He did so, and the flaw vanished.

Wollaston had well examined the stone, and, guided by his superior knowledge of the deceptive effects produced by the refractions of light that occur in crystals, and of the chance of dispersing them, his skill enabled him to succeed in the present instance. When the story got abroad, it became a question, among the frequenters of the clubs, whether the seller of the stone

ought not to have been apprised of the nature of the
flaw, and to have gone halves with the philosopher in
its improved value.

With all his learning and variety of knowledge, scientific and practical, Wollaston was not the most ready
to communicate it; nor was he always the most courteous when engaged in argument. He knew so much,
that he could not always avoid betraying a consciousness of his own superior knowledge, and of the want of
it in others. Still to those who understood him his
manner of disputing was not disagreeable. He would
rarely give an immediate or direct answer to a question,
but generally respond by putting another analogous
question, of an opposite tendency. Thus, for instance,
I ask him to define the word heat; he replies, "Tell
me how you define cold." I say, "I cannot." "Then
I will do it for you—cold is the absence of heat." But
having thus got him under weigh, a noble dissertation
on heat would surely follow.

In fact it was impossible to be in company with Dr.
Wollaston without acquiring new information, on whatever subject might be under discussion, more especially
when at the Club of the Royal Society, where the
members were intimately known to each other, and
great freedom of speech prevailed. Here it was not
unusual to start a subject for no other purpose than to
draw out Wollaston's opinion, or remarks upon it. I
must plead guilty of having not unfrequently done this.

In the 'Life of Sir Humphry Davy,' by Dr. Paris,
it is observed that "the chemical manipulations of
Wollaston and Davy offered a singular contrast to each
other, and might be considered as highly characteristic
of the temperaments and intellectual qualities of these

two remarkable men. Every process of the former was regulated with the most scrupulous regard to microscopic accuracy, and conducted with the utmost neatness of detail; while a degree of turbulence and apparent confusion attended the experiments of the latter; and yet each of them was equally excellent in his own style. By long discipline, Wollaston had acquired such power in commanding and fixing his attention upon minute objects, that he was able to recognise resemblances, and to distinguish differences, between precipitates produced by re-agents, which were invisible to ordinary observers. Davy, on the other hand, obtained his results by an intellectual process, which may be said to have consisted in the extreme rapidity with which he seized upon, and applied, appropriate means at appropriate moments."

To this faculty of minute observation which Dr. Wollaston employed with so much advantage, the chemical world is indebted for the introduction of more simple methods of experimenting, for the substitution of a few glass tubes and plates of glass for capacious retorts and receivers, and for the art of making grains give results, which previously required pounds. His laboratory therefore was a sealed recess to his most intimate friends.

On this subject Dr. Paris relates the story of a foreign philosopher who is said once to have called upon Dr. Wollaston with letters of introduction, and to have expressed a desire to see his laboratory. "Certainly," he said, and immediately produced a small tray, containing some glass tubes, a blow-pipe, two or three watch-glasses, a slip of platinum, and a few test-bottles: "*voilà mon laboratoire.*"

Wollaston, however, had no dislike to show by what small means he could produce great results. Having been engaged one day in inspecting a monster galvanic battery, constructed by Mr. Children, and having witnessed some of those powerful and brilliant phenomena of combustion which it was capable of producing, on his way home he accidentally met with a brother chemist, in the street, who was not unacquainted with Children's grand machine, and uttered something of the folly or inconvenience of such an enormous size; when Wollaston, seizing his button, led him into a bye corner, where, taking from his waistcoat pocket a tailor's thimble, which contained a galvanic arrangement, and pouring into it the contents of a small phial, he instantly astonished his friend by heating a platinum wire to a white heat. This certainly afforded a most remarkable contrast with the monster machine of Mr. Children.

About this time I took a favourable opportunity at the Club to beg that he would give me some notion of the process necessary to convert so insignificant an article as a thimble into a galvanic battery. " I will explain it to you," he said, "in a moment; nothing is more simple. Strike off the bottom, compress the two sides, and insert between them a small plate or bar of zinc, tin, or iron, and pour in a small quantity of sulphuric acid; it will become heated, and your galvanic battery is made." It may be observed that Wollaston was the first to demonstrate the identity of galvanism and common electricity.

Wollaston and Davy were among the chief contributors to the ' Philosophical Transactions,' and Mr. Weld, in his recent ' History of the Royal Society,' says that, in this collection, Dr. Wollaston's papers amount to thirty-nine, in addition to those on strictly chemical subjects; that they include memoirs on Astronomy, Optics, Mechanics, Acoustics, Mineralogy, Crystallography, Physiology, Pathology, and Botany. To enumerate his various inventions would require more space than could be afforded in this memoir. It has been said that Wollaston's knowledge was more varied, and his taste less exclusive, than that of any other philosopher of his time, with the exception perhaps of Mr. Cavendish; but apart from science, according to Mr. Weld, little remains to be told.

He observes, "No picturesque incidents nor romantic stories adorn Wollaston's biography, and but few characteristic anecdotes have been preserved. His days were spent between his laboratory and his library, his evenings at the meetings of the Royal, the Geological, and other Societies, in whose proceedings he always took a keen interest." He made, however, occasional excursions into the country; and we are told by Sir Humphry Davy that "this illustrious philosopher was nearly of the age of fifty before he made angling a pursuit; yet he became a distinguished fly-fisher; and the amusement occupied many of his leisure hours during the last twelve years of his life."

The knowledge of Wollaston, however, was of too much public importance to allow the Government to overlook it. Accordingly in 1814 they made an application to the Royal Society to appoint a committee of the most eminent men of science to examine certain gas-works in the metropolis, with the view of ascertaining whether they were efficiently constructed, Wollaston being the principal member. A report was drawn up to the effect that " there was a great necessity for the improvement of some of the works, and a propriety of occasional superintendence of all of them."

In 1818 Mr. Croker, then First Secretary to the Admiralty, introduced and passed an Act of Parliament for a new constitution of the Board of Longitude, by which three Fellows of the Royal Society were added to the Board, as Commissioners, with salaries of £100 a year each, who were to be a kind of Council to the Admiralty on scientific subjects. The first three commissioners were Dr. Wollaston, Dr. Young, and Captain Kater, but subsequently Dr. Young was appointed Secretary, with the charge of the supervision of the ' Nautical Almanac,' to which a salary of £300 a year was attached, and Colonel Mudge succeeded him as commissioner. This was the first attempt, that I remember, to open salaried office to men of science; and small as the boon was, it was gratefully received; and the system worked with great advantage both to the Admiralty and the public. But, in a subsequent fit of economy, this poor pittance was withdrawn from science.

In 1819 Dr. Wollaston was appointed a member of two Committees: one, " For Inquiry into the mode of preventing the Forgery of Bank Notes," consisting of

Sir Joseph Banks,	Dr. Wollaston,
Davies Gilbert, M.P.	Mr. Hatchett,

And two others;

the other, "To consider the subject of Weights and Measures "—

Sir Joseph Banks.	Dr. Wollaston.
Sir George Clark, M.P.	Dr. Young.
Davies Gilbert, M.P.	Capt. Kater.

The independent spirit of Dr. Wollaston was manifested on many occasions, and it prevented him from
making solicitations for favours, however trifling, from
men in power, feeling a strong repugnance to subject
himself to a refusal, or, if granted, to the inconvenience
of laying himself under obligations which would, to a
certain degree, trench upon his independence. A copy
of one of his letters was put into my hands, so characteristic of his feelings on this subject, and evincing so
noble a trait of generosity, that I cannot forbear to
insert it, as I have permission to do.

Dr. Wollaston to his Brother, Mr. Henry Wollaston.

" My dear Henry,

" I have long been prepared to prove how truly
I respect your conduct through life, from first to last,
and how willing I am to assist you in a way that
I can.

" I wish it were in my power to procure for you the
situation in the Customs you wish me to apply for, or
any other where your talents, assiduity, and prudent
management might be turned to certain account; but I
decline making solicitations with probability of gaining
nothing, and with absolute certainty of forfeiting a portion of that independence on which my happiness in life
depends.

" By the transfer which I enclose I do not deprive
myself of any of those comforts or even indulgences to
which I think myself entitled for a certain amount of
continued exertion and steady economy; and I shall
live with the satisfaction of having disposed of a part of

my property to the best account, instead of reflecting that I shall not live long enough to have occasion for it.

              " Believe me ever affectionately
                      " and sincerely yours,

                         " W. H. WOLLASTON."

The enclosed, alluded to, was a stock receipt for *ten thousand pounds*, 3 per Cent. Reduced.

But Wollaston's generosity was not merely confined to his family relations. He was not unmindful of the source from whence his fame and fortune were derived. Under the presidency of the Royal Society, Mr. Davies Gilbert holding the chair, Dr. Wollaston established, for the promotion of science, what was thence called *the Donation Fund*. At a meeting of Council held on the 11th of December, 1828, the following communication from Dr. Wollaston was read:—

          " No. 1, Dorset Street, 26th Nov. 1828.

    " I have this day vested 2000*l*. 3 per Cent. Consolidated Bank Annuities in the joint names of myself and the President, Council, and Fellows of the Royal Society of London, in trust, that the said trustees shall, during my life, pay to me the dividends on the said stock; and that, after my decease, the President, Council, and Fellows aforesaid shall apply the said dividends, from time to time, in promoting experimental researches, or in rewarding those by whom such researches may have been made; or in such manner as may appear to the President and Council for the time being most conducive to the interests of the said Society in particular, or of science in general; which latter application of the said dividends will, in my opinion, be most creditable to the President and Council."

And he adds—

" I hereby enjoin the said President, Council, and Fellows, after my decease, not to hoard the said dividends parsimoniously ; but to expend them liberally, as nearly as may be annually, in furtherance of the above-declared objects of the trust.

" W. HYDE WOLLASTON."

At the same time the Council were informed that Mr. Davies Gilbert had contributed 1000*l.* to the above-mentioned fund, Mr. Charles Hatchett 105*l.*, and that several others had subscribed. Dr. Wollaston more-over left a legacy of 1000*l.* to the Geological Society, to found a Donation Fund.

Towards the latter part of 1828 Dr. Wollaston became dangerously ill with a disease of the brain. Finding his end approaching, and being personally unable to write out accounts of such of his discoveries and inventions as he was desirous should not perish with him, he employed an amanuensis, and in this manner communicated some of his most valuable papers to tne Royal Society.

In December his illness increased rapidly, accompanied with severe suffering; but it may be stated, as an interesting fact, that, in spite of the extensive cerebral disease under which he laboured, his faculties were unclouded to the end. When he was nearly in the last agonies, one of his friends having observed, loud enough for him to hear, that he was not conscious of what was

F

passing around him, he made a sign for pencil and
paper. He then wrote down some figures, and, after
casting up the sum, returned the paper. The amount
was correct. He died on the 22nd of December, 1828,
aged sixty-two, only a few months before his great
scientific contemporaries, Sir Humphry Davy and Dr.
Thomas Young.

It is somewhat remarkable, and much to be regretted,
that so acute a philosopher, and so distinguished a
chemist and mechanist, enjoying so high a situation
among the learned, should, like Sir Joseph Banks, have
found no biographer to record his merits and transmit
his fame to posterity.

Mr. Weld has given a short character of Wollaston,
extracted from Dr. Henry's 'Chemistry.' "He was
remarkable for the caution with which he advanced from
facts to general conclusions; a caution which, if it sometimes prevented him from reaching at once to the most
sublime truths, yet rendered every step of his ascent a
secure station, from which it was easy to rise to higher
and more enlarged inductions."

"Wollaston," says Mr. Brande, "was a good chemist, but his papers are chiefly on physical and physio-
logical inquiries; he was remarkable for the singular and
satisfactory simplicity of his experimental methods, for
the perspicuity of his theoretical deductions, and for the
extreme caution with which he touches upon generalizations and hypotheses."

"In 1827 he first felt a numbness in his left arm,
which he considered to be a symptom of paralysis, to
which his father and eldest brother had fallen victims,
and after them a second brother. The following year he
was again seized when on a fishing excursion at Stock-

bridge ; and in the same year, when on a visit to Lord
Spencer in the Isle of Wight, the retina of the left eye
became insensible to the action of light.   He returned
home and became better; but soon after the sensation
in his arm became worse, and in the course of a week
symptoms more decidedly alarming came on : the use of
his arm was much impaired, and the muscles of the face
and organs of speech were affected ; his mental faculties were, however, entire to the last."

Sir Humphry Davy has added his tribute to the
memory of Dr. Wollaston, but he has contrived to
spoil it, by an ungenerous observation with respect to
his application of science to profit—without adverting
to the profit that science had so abundantly poured
into his own lap.

"Dr. W. Hyde Wollaston," he says, "may be compared with Dalton for originality of view, and even for
his superior accuracy.   He was an admirable manipu-
lator, steady, cautious, and sure; his judgment was
cool, his views sagacious, his inductions made with
care, strongly formed, and seldom renounced.   He had
much of the same spirit of philosophy as Cavendish;
but, unlike Cavendish, he applied science to purposes of
profit, and for many years sold manufactured platinum.
He died very rich.   Some accidental annoyances in the
medical profession made him, I think, jealous and
reserved in the earlier part of his life; but latterly he
became far more agreeable and confiding, and was a
warm and kind friend, and a pleasant social companion."

The allusion to Mr. Cavendish's supposed liberality
is singularly unjust to Dr. Wollaston.   Mr. Cavendish,
as we shall see presently, was by inheritance, and
saving, immensely rich.   He had no occasion to look

to science for profit—nor do I know that any of his discoveries were capable of being turned to that sort of account.

On the anniversary of November, 1828, two medals were given — one of them to Dr. Wollaston, then labouring under the dangerous illness which terminated in death.   On this occasion the President, Mr. Davies Gilbert, thus alluded to those circumstances :—

" The other royal medal has been awarded by your Council for a communication made under circumstances the most interesting and most distressing.   On the first day of our meeting, a paper from Dr. Wollaston was read, descriptive of the processes and manipulations by which he has been enabled to supply all men of science with the most important among the recently discovered metals, platinum—possessing various qualities useful in an eminent degree to chemists, even on a large scale, and resisting fusion in the most intense heat of our wind furnaces.   Alloyed indeed with arsenic, it became susceptible of receiving ornamental forms ; but a continued heat expelled the volatile metal, and left the other in a state wholly unfit for use.   Dr. Wollaston, instead of alloying, purified the platinum from every admixture, by solution, consolidated its precipitate, by pressure, by heating, and by percussion, so as to effect a complete welding of the mass, thus being made capable of being rolled into leaf, or drawn into a wire of a tenacity intermediate between those of iron and gold.   To these scientific and beautiful contrivances we owe the use of a material, not only of high importance to refined chemistry, but now actually employed, in the largest manufactories, for distilling an article of commerce so abundant and so cheap as sulphuric acid.   And, above all, we owe to them the material which, in the skilful hands of some members of this society, has mainly contributed to their producing a new species of glass, which promises to form an epoch in the history of optics. The Council have therefore deemed themselves bound to express their strong approbation of this interesting memoir, by awarding a royal medal to its author."

The following papers appeared from the pen of Wollaston from 1822 to 1830 :—

1. On the concentric adjustment of a triple object-glass.

2. On the finite extent of the atmosphere.

3. On metallic titaneum.

4. On the apparent magnetism of metallic titaneum.

5. On the semi-decussation of the optic nerves.

6. On apparent direction of eyes in a portrait.

7. Bakerian Lecture on a method of rendering platinum malleable.

8. Description of a microscopic doublet.

9. On a method of comparing the light of the sun with that of the fixed stars.

10. On the water of the Mediterranean.

11. On a differential barometer.

12. The cutting diamond.

The whole amount of Dr. Wollaston's papers in the 'Philosophical Transactions' are stated by Mr. Weld at thirty-nine.

As a specimen of the manner in which he describes an article of his invention, *a Method of Freezing at a Distance*, may be taken.

He commences by giving an ingenious plan, of Mr.

Leslie, for employing an extensive surface of sulphuric acid, to absorb the vapour generated in the course of the experiment.

" But," he says, " in this method the labour is not inconsiderable, and the apparatus, though admirably adapted to the purpose for which it is designed, is large and costly. I have therefore thought the little instrument I am about to describe may possess some interest, as affording a readier and more simple mode of exhibiting so amusing and instructive an experiment.

" Let a glass tube be taken, having its internal diameter about one-eighth of an inch, with a ball at each extremity of about one inch diameter; and let the tube be bent to a right angle at the distance of half an inch from each ball. One of these balls should contain a little water (less than half full), and the remaining cavity should be as perfect a vacuum as can readily be obtained. One of the balls is made to terminate in a capillary tube; and when water admitted into the other has been boiled over a lamp for a considerable time, till all the air is expelled, the capillary extremity, through which the steam is still issuing with violence, is held in the flame of the lamp till the force of the vapour is so far reduced that the heat of the flame has power to seal it hermetically.

" When an instrument of this description has been successfully exhausted, if the ball that is empty be immersed in a freezing mixture of salt and snow, the water in the other ball, though at the distance of two or three feet, will be frozen solid in the course of a few minutes. The vapour contained in the empty ball is condensed by the common operation of cold, and the vacuum produced by this condensation gives opportunity for a fresh quantity to arise from the opposite ball, with proportional reduction of its temperature.

" According to a theory that does not admit of positive cold, we should represent the heat of the warmer ball to be the agent in this experiment, generating steam as long as there remains any excess of heat to be conveyed. But if we would express the cause of its abstraction, we must say that the cold mixture is the agent, and may observe, in this instance, that its power of freezing is transferred to a distance by what may be called the negative operation of steam.

" The instrument by which this is effected may aptly be called a cryophorus, which correctly expresses its office of frost-bearer."

Section IV.

Sir Humphry Davy.

Humphry Davy was born at a village near Penzance, 17th of December, 1778.   His father, Robert Davy, was a carver in wood, and known, from his diminutive size, as " the little carver;" a man of some small property in land, and not deficient in his craft.   It is said that at five years of age Humphry would turn over the pages of a book as rapidly as if counting the number of leaves or hunting after pictures, yet, on being questioned, could give a satisfactory account of its contents; a faculty which, Lady Davy has said, was retained by him through life: so enduring are early impressions and habits even to old age.

' The Pilgrim's Progress' was the book that first attracted his attention.   At eight years of age he was a great lover of the marvellous, and composed stories of romance and tales of chivalry; thus early commenced his fondness for fiction.   He was also in the habit of writing verses and ballads.   He was placed at a very early age in a grammar-school at Penzance, and lived with Mr. Tonkin, a kind friend of the family; but in the holidays Humphry always resided with his parents, on the small property they possessed at Varfell.

He was extremely fond of fishing, and, when strong enough to carry a gun, much of his leisure time was passed in shooting birds, which he is said to have stuffed and set up with more than ordinary skill.

Davy on one occasion got up a pantomime, and his biographer has given the *dramatis personæ*, with the actors' names, which he found on a fly-leaf torn out of a Schrevelius' Lexicon.

In 1793 from Penzance he went and finished his school-education under the Rev. Dr. Cardew, a preceptor distinguished by the number of eminent scholars whom he sent forth to the world. By Dr. Cardew's account to Mr. Davies Gilbert, Humphry Davy was not one of them. He says, " With respect to our illustrious countryman, Sir H. Davy, I fear I can claim but little merit from the share I had in his education. He was not long with me; and while he remained I could not discern the faculties by which he was afterwards so much distinguished; I discovered indeed his taste for poetry, which I did not omit to encourage."

His temper during youth is represented as mild and amiable. He never suppressed his feelings, but every action was marked by ingenuousness and candour, qualities which endeared him to his associates, and gained him the love of all who knew him. In 1794 his father died; and the widow, having taken up her residence in Penzance, apprenticed her son, by the advice of her long-valued friend Mr. Tonkin, to Mr. John Bingham Borlase, a surgeon. But Davy's mind, having for some time been engrossed with philosophical inquiries, was bent upon the pursuit of his own plans of study; the utensils of the shop and the kitchen were his chemical apparatus.

It was some time after he had been placed with Mr Borlase that he indicated any turn for chemistry, the

study of which he then commenced with all the ardour
of his temperament. It was first directed by a desire to
discover various mixtures as pigments, a suggestion to
which, however, his biographer seems not to pay much
attention.

The Rev. Dr. Batten, the Principal of the East India
College at Hayleybury, paid a visit to his friend in
Cornwall, where he had been one of Davy's earliest
schoolfellows. In the course of conversation Dr. Batten
spoke of the attention he had been paying to the principles of mechanics, and expressed himself more particularly pleased with that part which treats of the 'Col-
lision of Bodies.' What was his surprise on finding
Davy as well, if not better, acquainted with its several
propositions! Yet he had never systematically studied
the subject—had never perhaps seen any standard work
upon it; but he had instituted experiments with elastic
and inelastic balls, and had worked out the results by
the unassisted energies of his own mind. It appeared
clear that, had this branch of science not existed, Davy
would have created it.

In his evening walks along the beach his constant
companion was a hammer to knock off specimens from
the rocks. "In short," says his biographer, "at this
period he paid much more attention to philosophy than
to physic; he thought more of the bowels of the
earth than of the stomachs of his patients; and
when he should have been bleeding the sick, he was
opening veins in the granite."

Mr. Tonkin's garret had now become the scene of
his chemical operations; and upon more than one occasion it is said that he produced an explosion which put
the Doctor and all his glass bottles in jeopardy. "This

boy Humphry is incorrigible! was there ever so idle a
dog!—he will blow us all into the air!" Such were the
constant exclamations of Mr. Tonkin; and then, in a
jocose strain, he would talk of him as the "Philosopher," and sometimes call him " Sir Humphry," as if
prophetic of his future renown.

It was Davy's great delight, as I have said, to ramble
along the sea-shore, and often, like the orator of Athens,
would he on such occasions declaim against the howling
of the wind and waves, with a view to overcome a defect
in his voice, which was but slightly perceptible in his
maturer age. It is said that the peculiar intonation he
employed in his public addresses, and which rendered
him obnoxious to the charge of affectation, was the result
of a laborious effort to conceal this natural infirmity.

That in his youth Davy possessed courage and deci-
sion may be inferred from the circumstance of his
having, upon receiving a bite from a dog supposed to
be rabid, taken his pocket-knife, and, without the least
hesitation, cut out the part on the spot, and then
retired into the surgery and cauterized the wound; an
operation which confined him to Mr. Tonkin's house
three weeks. This anecdote of Davy's excision of the
part, with so much promptitude and coolness, derives
an interest from the age and inexperience of the
operator.

A French vessel having been wrecked off the Land's
End, the surgeon escaped, and made his way to Penzance. Davy met him by accident and showed him
many civilities, and received in return a case of instruments which had been saved from the ship. They were
eagerly examined by the young chemist to see how far
they might be convertible to experimental purposes.

A clyster-machine was viewed with exultation, and
in the course of an hour did this machine, emerging
from its insignificance, figure away in all the pomp and
glory of a complicated piece of pneumatic apparatus;
and at length we are told it actually performed the
duties of an air-pump, in an original experiment on the
nature and sources of heat, which was the subject of
discussion and experiment shortly after, by Black and
some other philosophers.

That the romantic scenery of Cornwall, with its mine-
ralized rocks, should have tempted a youthful genius,
living among friends and associates connected with
mining speculations, to turn his mind to the study of
geology, chemistry, and mineralogy, is nothing more
than might have been expected, nor could such scenery,
in the mind of one possessing a quick sensibility to the
sublime forms of nature, fail to kindle that enthusiasm
which is so essential to poetical genius; accordingly
we learn that Davy became enamoured of the Muses at
a very early age, and evinced his passion by several
poetical productions, some of which are introduced in
his ' Life.'

His biographer observes, that, as these specimens have
now become scarce, he has thought it right to place them
on record, as bearing the stamp of lofty genius; a vein
of philosophical contemplation running through their
composition, and an ardent aspiration after fame. But,
he adds, there is a higher motive, " that of observing
the bias of his genius at this early period, with a view
to compare it with that which displayed itself in the
last days of the philosopher." We shall find that the
bright and rosy hues of fancy which gilded the morning
of his life were subdued or chased away, but that they

again glowed forth in the evening of his days, and illu-
mined the setting, as they had the dawning of his genius.

The following anecdote shows Davy's progress in
early life.—Mr. Davies Gilbert, walking with a friend,
observed a boy, with a comely countenance, carelessly
swinging on a gate, and took notice of his intelligent
expression. " That boy," said the friend, " is young
Davy, the carver's son, who," he added, " is said to
be fond of making chemical experiments." " Chemical experiments!" exclaimed Mr. Gilbert, with much
surprise ; " if that be the case, I must have some
conversation with him." He soon discovered ample
evidence of the boy's singular genius ; offered young
Humphry the use of his library, or any other assistance he might require for the pursuit of his studies ;
and gave him an invitation to his house at Treadrea,
of which Davy frequently availed himself.

Dr. Beddoes, having established the Pneumatic
Institution at Bristol, required an assistant who might
superintend the necessary experiments in the laboratory; and Mr. Gilbert proposed Davy as a person
fully competent to fill the situation. Beddoes readily
availed himself of Gilbert's suggestion, saying, "I am
glad that Davy has impressed you as he has me;" and
he wished him to arrange with regard to the salary,
observing that the fund would not furnish a salary from
which a man could lay up anything. In a second letter
Dr. Beddoes says, " I have received a letter from Mr.
Davy in which he has oftener than once mentioned a
genteel maintenance as a preliminary to his being employed to superintend the Pneumatic Hospital."

Mr. Gilbert kindly undertook the negotiation, and
completed it to the satisfaction of all the principal

parties. Mrs. Davy yielded to her son's wishes, and
Mr. Borlase very generously surrendered his indenture,
with an endorsement to the following effect—" That
he freely gave up the indenture, on account of the
singularly promising talents which Mr. Davy had displayed." On the 2nd of October, 1798, Davy quitted
Penzance, before he had attained his twentieth year.
Mr. Gilbert well remembered meeting him upon his
journey to Bristol, in the highest spirits, and in that
frame of mind in which a man of ardent imagination
identifies every successful occurrence with his own for-
tunes; Davy's exhilaration, therefore, was not a little
heightened by the arrival of the mail-coach from London, covered with laurels and ribbons, and bringing
the news, so cheering to every English heart, of Nelson's glorious victory of the Nile.

Davy now gave the reins to a flood of philosophical
theories, which he very soon discovered to be what he
himself calls "the dreams of misemployed genius," and in
due time regrets that he had ever published the essays that
contain them; but, as his biographer Dr. Paris observes,
the reader will be disposed to treat them with all tenderness, when he considers that the author of them was
barely eighteen years of age; and, moreover, Beddoes
sanctioned the publication: much as Davy needed the
bridle, Beddoes required it still more. The only pun
that Davy is said to have ever made was on the occasion of Mr. Sadler being appointed by Dr. Beddoes as
his (Davy's) successor, when he remarked, " I cannot
imagine why he has engaged *Sadler*, unless it is that
he may be well *bridled*."

Several very interesting letters to Mr. Davies Gilbert
are inserted by his biographer, containing a series of

chemical experiments, chiefly on the gases, two of which, exercised upon himself, had very nearly deprived him of life.   His *Researches* however, unlike his *Essays*, seem to have excited very general admiration in the philosophic world; and fortunately for Davy, while the vivid impression produced by his new work was in full glow, Count Rumford was anxiously looking for some rising philosopher who might contribute his energies towards the promotion of that fame which the recently established " INSTITUTION OF GREAT BRITAIN " had acquired in chemistry—an institution that has pro-gressed to its state of celebrity, chiefly, by the talents and the exertions of such men as Davy, Brande, and Faraday.

Davy was appointed; and the minute of the managers, dated 16th of February, 1801, runs thus:—

" Resolved, that Mr. Humphry Davy be engaged in the service of the Royal Institution, in the capacities of the Assistant Lecturer in Chemistry, Director of the Laboratory, and Assistant Editor of the Journals of the Institution; and that he be allowed to occupy a room in the house, and be furnished with coals and candles; and that he be paid a salary of one hundred guineas per annum."

His biographer relates, as a curious fact, that the first impression produced on Count Rumford by Davy's personal appearance was highly unfavourable to the young philosopher, and he expressed to Mr. Underwood his great regret at having been influenced by the ardour with which his suit had been urged; and he actually would not allow him to lecture in the theatre until he had given a specimen of his abilities in the smaller lecture-room.   His first lecture, however, entirely removed every prejudice which had been formed ; so that at its conclusion the Count emphatically exclaimed, " Let him command any arrangements which the Institution can afford." He was accordingly, on the very next day, promoted to the great theatre.

It is further said that " Davy's uncouth appearance and address subjected him to many other mortifications on his first arrival in London. There was a smirk on his countenance, and a pertness in his manner, which, although arising from the perfect simplicity of his mind, were considered as indicating an unbecoming confidence." Johnson the publisher was in the custom of giving weekly dinners to the more distinguished authors and literary stars of the day. Davy, soon after his appointment, was invited upon one of these occasions, but the host actually thought it necessary to explain to his guests, by way of apology, the motives which had induced him to introduce into their society a person of such humble pretensions. At this dinner, a circumstance occurred which must have been very mortifying to the young philosopher. Fuseli was present, and, as usual, highly energetic upon various passages of beauty in the poets, when Davy unfortunately observed, that there were passages in Milton which he could never understand. " Very likely, very likely, Sir," replied the artist, in his broad German accent, " but I am sure that is not Milton's fault."

He performed the duties of his station at the Royal Institution so greatly to the satisfaction of the managers that, six weeks after his first entrance, the following resolutions were passed :—

" Resolved, that Mr. Humphry Davy, Director of the Chemical Laboratory, and Assistant Lecturer in

Chemistry, has, since he has been employed at the Institution, given satisfactory proofs of his talents as a lecturer.

" Resolved, that he be appointed, and in future denominated, Lecturer in Chemistry at the Royal Institution, instead of continuing to occupy the place of *Assistant* Lecturer, which he has hitherto filled."

An early friend of his, who says, " I loved him living —I lament his early death—I shall ever honour his memory," has given an account of the impression which Davy's first course of lectures made on the public :—

" The sensation created by his first course of lectures at the Institution, and the enthusiastic admiration which he obtained, is at this period scarcely to be imagined. Men of the first rank and talent, the literary and the scientific, the practical and the theoretical, blue stockings and women of fashion, the old and the young, all crowded —eagerly crowded the lecture-room. His youth, his simplicity, his natural eloquence, his chemical know-ledge, his happy illustrations, and well-conducted expe-riments, excited universal attention and unbounded applause. Compliments, invitations, and presents were showered upon him in abundance from all quarters ; his society was courted by all, and all appeared proud of his acquaintance."

At this point of his biography Dr. Paris observes, " I should not redeem the pledge given to my readers, nor fulfil the duties of an impartial biographer, were I to omit acknowledging that the manners and habits of Davy very shortly underwent a considerable change. Let those who have vainly sought to disparage his excellence enjoy the triumph of knowing that he was not perfect ; but it may be asked in candour, where is

G

the man of twenty-two years of age, unless the temperament of his blood were below zero, and as dull and pas-
sionless as the fabled god of the Brahmins, who could
remain uninfluenced by such an elevation ?"

That such a change should have taken place is
hardly to be wondered at. That his vanity should be
excited, his ambition raised, and the simplicity of his
manners should lose its bloom, amidst the buzz of adula-
tion, are results to have been expected—nor was a
moment left to him for reflection. In short, so popular
did he become, under the auspices of the Duchess of
Gordon and other leaders of high fashion, that even
their *soirées* were considered incomplete without his
presence; and yet these fascinations, strong as they
must have been, never tempted him from his allegiance
to science; never did the charms of the saloon allure
him from the pursuits of the laboratory, or distract him
from the duties of the lecture-room. The crowds that
repaired to the Institution in the morning were, day
after day, gratified by newly devised and highly illustrative experiments, conducted with the utmost address,
and explained in language at once perspicuous and
eloquent.

His biographer adds, "Had Davy, at this period
of his life, been anxious to obtain wealth, such was his
reputation in chemistry, and such the value attached to
his judgment, that, by lending his assistance to manufac-
turers and projectors, he might easily have realised it;
but his aspirations were of a nobler kind—Scientific
Glory was the grand object for which his heart panted;
by stopping to collect the golden apples he might have
lost the race."

It was a general observation among his friends, how-

ever, as I have just observed, that, by the successful
result of his lectures, his manners had undergone a
change; and that, to their regret, he had lost much of
his native simplicity. To the anxiety of Mr. Poole, who
apparently had touched upon this point, Davy says,
" Be not alarmed, my dear friend, as to the effect of
worldly society on my mind. The age of danger has
passed away. There are, in the intellectual being of all
men, permanent elements, certain habits and passions
that cannot change. My *real*, my *waking* existence, is
amongst the objects of scientific research; common
amusements and enjoyments are necessary to me only
as dreams to interrupt the flow of thoughts too nearly
analogous to enlighten and to vivify."

On the death of Dr. Gray, Secretary of the Royal
Society, Mr. Davy was appointed his successor, in
January, 1807, when he was also elected a member of
the Council.

The great variety of discoveries made, in rapid succession, by Davy raised his renown in every part of
Europe. In France the prize founded by Buonaparte,
for the encouragement of electric researches, was awarded
to him. His intense application to objects of philosophical research produced, in the year 1807, a severe illness.
But he recovered and returned with invigorated ardour
to the prosecution of his philosophical investigations,
so that the fame of Davy had, before the year 1810, been
complete, and the high importance of his discoveries
had become the general theme of admiration, throughout the scientific societies of Europe. In the latter
year the members of the Dublin Society were induced
to invite him to that city for the purpose of delivering
a course of lectures. Davy readily complied with the

request; and, when the lectures were completed, a resolution was passed, "That the thanks of the Society be
communicated to Mr. Davy for the excellent course
of lectures," &c.; and a request made that he would
accept the sum of five hundred guineas from the
Society.

In the following year he was again solicited to deliver
lectures in their laboratory, to which he assented, and
delivered two distinct courses—one on the Elements
of Chemical Philosophy, and the other on Geology;
for which he received the unanimous thanks of the
Society, and a letter enclosing a draft for seven hundred
and fifty pounds.

On his return he was requested by Lord Liverpool
to meet him at the House of Lords, to consider of a
mode of ventilating it. Davy, however, admitted the
experiment he made to have been a complete failure.
Vexatious as this was to Davy, it became of course to
others a fertile source of pleasantry, and gave rise to
numerous epigrams, which, in recording the miscarriage
of science, endeavoured to display the triumph of wit.
It did not, however, prevent the Prince Regent from
bestowing on Davy the honour of Knighthood, at a
levee held at Carlton House on the 8th of April, 1812.

The following day Sir Humphry delivered his farewell lecture before the members of the Royal Institution,
for he was on the eve of assuming a new station in
society, which decided him to retire from those public
situations which he had long held with so much advantage to the world, and with so much honour to himself.
Two days after this, Sir Humphry Davy was married
to Mrs. Apreece, the widow of Shuckburgh Ashby
Apreece, Esq., eldest son of Sir Thomas Apreece.

This lady was the daughter and heiress of Charles Kerr, Esq , of Kelso, and possessed a very considerable fortune.

On his journey to the North in August, 1812, he thus writes to his friend Purkis:—" Receive my warm acknowledgments for your kind congratulations on my becoming a Benedick. I can now speak from experience, in which you have long participated: I am convinced that the natural state of domestic society is the best fitted for man, whether he be devoted to philosophy or to active life." In October he says to Children, " I shall be ready to come to my business whenever you think I can be useful. I shall not be able to endure a very long separation from my wife, but for three or four days I am at your command."

He had enough of business before him. His investigations and experiments, made on the various acids and the gases, employed much of his time, besides the preparation for publishing a work on the 'Elements of Agricultural Chemistry.' In October, 1813, his attention was turned towards making a visit to Paris, and, having obtained the permission of Buonaparte (granted to very few), he started on his journey; accompanied by Lady Davy, and by Mr. Faraday as secretary and chemical assistant.

Early in the above year Mr. Faraday, by the recommendation of Sir Humphry Davy, had been appointed to the situation of assistant in the laboratory of the Royal Institution. He says, in a letter to Dr. Paris— " Through the good efforts of Sir Humphry, I went to the Royal Institution early in March, 1813, as assistant in the laboratory; and in October of the same year went with him abroad as his assistant in experiments and in writing. I returned with him in April, 1815, resumed my station in the Royal Institution, and have ever since remained there." And may he long continue there to enlighten and delight his numerous and intelligent audiences !

On the party landing at Morlaix they were arrested by the local authorities, who, notwithstanding their passport, thought it impossible that a party of English could, under any circumstances, have obtained permission to travel over the Continent. They were therefore compelled to remain in the town of Morlaix six or seven days, until the necessary instructions could be received from Paris. On the 27th they reached the French capital.

When Sir Humphry visited the Louvre, he walked with a rapid step along the gallery, and, to the astonishment of the *cicerone*, did not direct his attention to a single painting. The only exclamation of surprise that escaped him was, " What an extraordinary collection of fine frames ! " When opposite to Raphael's picture of the Transfiguration, his conductor, in a tone of enthusiasm, desired the philosopher's attention to that most sublime production of art, when Davy's reply was laconic as it was chilling—" Indeed ; I am glad I have seen it ;" and then hurried forward, as if desirous of escaping from any critical remarks upon its excellence.

The statues in the lower department were regarded by Davy with the same kind of frigid indifference— " while the marble glowed with more than human passion, the living man was colder than the stone." The apathy, the total want of feeling he betrayed, on having his attention directed to the Apollo Belvidere,

the Laocoon, and the Venus de Medicis, was as inexpli-
cable as it was provoking; but an exclamation of the
most vivid surprise escaped him at the sight of an
Antinous, treated in the Egyptian style, and sculptured
in *alabaster*: "Gracious powers," said he, " what a
beautiful stalactyte!"

His whole conduct, in fact, while in Paris, was most
unaccountable, whether with regard to things or to persons. The colossal elephant, which was intended to
form a part of the fountain then erecting on the site of
the Bastile, caught Davy's fancy more than any object
he there met with.

Davy, while in Paris, incurred the displeasure of
some of the chemists by his undue interference, as they
considered it, with a new material which they had discovered, and the properties of which he had investi-
gated.   " While great bitterness of feeling towards
Davy existed by Thénard and Gay-Lussac, on account
of the affair of iodine, Chevreul and Ampère are, and
were, of opinion that it had its origin in a misconcep-
tion; that what Davy did was from the honest desire
of promoting science, and not from any wish to detract
from the merit of the French chemists." Davy most
unquestionably worked out, and in the French capital
too, the true nature of iodine, and found it to be a new
and peculiar acid, and that it was a substance analogous
in its chemical relations to chlorine; he should have
made a direct communication to the original discoverers.

Sir Humphry never saw the Emperor while at Paris,
and objected to attend his levee; but he and Lady
Davy were presented to the Empress at Malmaison.
In December, Davy was elected a corresponding member

of the first class of the Imperial Institute; and nothing
indeed could have exceeded the liberality and unaffected kindness with which the *savans* of France
received and caressed the English philosopher.

Sir Humphry's biographer says it would be an act
of literary dishonesty to assert that he returned the
kindness of the *savans* of France in a manner that the
friends of science could have expected and desired; that
there was a flippancy in his manner, a superciliousness
and hauteur in his deportment, which surprised as much
as they offended.

But in thus avowing the errors of Davy, he claims
from the readers of his Life credit for the sincerity with
which he attempts to palliate them. " From my personal knowledge of his character, I am inclined to refer
much of that unfortunate manner, which has been considered as the expression of a haughty consciousness of
superiority, to the desire of concealing a *mauvaise honte*
and *gaucherie*—an ungraceful timidity, which he could
never conquer."

A short time after his return to England he received
an invitation from the Rev. Dr. Grey, afterwards Bishop
of Bristol, calling his attention to the numerous and
awful accidents that were constantly occurring from the
explosion of inflammable air, or *fire-damp*, in the coal-
mines of the north of England, the remedy for which
had hitherto defied the skill of the best practical en-
gineers and mechanics of the kingdom, in consequence
of which, it appears from a paper in the Philosophical
Transactions of 1813, " in the space of seven years upwards of three hundred pitmen had been suddenly
deprived of their lives, besides a considerable number
who had been severely wounded; and that more than

three hundred women and children had been left in a
state of the greatest distress and poverty."

In August, 1815, Sir Humphry Davy answered Dr.
Grey's letter, by thanking him for calling his attention
to so important a subject, and stated that it would give
him great satisfaction if his chemical knowledge could
be of any use in an inquiry so interesting to humanity,
and that if he thought his inspecting the mines might be
useful he would cheerfully do so. He visited Dr. Grey
at Bishop-Wearmouth; was introduced to Mr. Buddle,
an intelligent gentleman, of the highest authority on all
subjects connected with the art of mining, from whom he
received important information; and in October, 1815,
he wrote to Dr. Grey, from the Royal Institution, telling him that the results of his investigations of the *fire-
damp* had been successful, far beyond his expectations,
and that he had never received so much pleasure from the
result of any of his chemical labours; "For I trust," he
said, "the cause of humanity will gain something by it."

On the 9th of November, 1815, Sir Humphry sent
a paper to the Royal Society, labelled, ' On the Fire-
Damp of Coal-Mines, and on Methods of Lighting the
Mine so as to prevent its Explosion.' In this paper
will be found the whole theory and operation of the
safety-lamp.

On the 15th of December, 1815, he thus wrote to
Dr. Grey :—

" I shall forward my lanterns and lamps to you in a
few days. They are *absolutely* safe; and if the miners
have any more explosions from their light, it will be
their own fault." In a further letter he says, " I have
made very simple and economical lanterns and candle-
guards, which are not only *absolutely safe,* but which

give light by means of the *fire-damp*, and which, while they disarm this destructive agent, make it useful to the miner."

This discovery of the *safety-lamp*, and all others to which it gave rise, whether considered in relation to their scientific importance, or to their great practical value, must be regarded among the most splendid triumphs of human genius.

Committees of coal-owners, proprietors of coal-mines, and the miners themselves, were all anxious that some general tribute of respect and gratitude should be paid to the inventor of the safety-lamp, so admirably calculated to obviate those dreadful calamities, and that lamentable sacrifice of human life, which had so frequently occurred in the mines of the north of England. A service of plate was presented to him from the coalowners of the Tyne and Wear, at a public dinner, as a testimony of their gratitude for the services he had rendered to them, and to humanity.

There was, however, another candidate aspiring to credit for the invention of the safety-lamp, Mr. George Stephenson, whose claims were supported by certain coal-owners and highly respected gentlemen of the North, at the head of whose names stood that of the Earl of Strathmore.

In order to repel the assertions put forward in certain resolutions of this party, a respectable body of coalowners had a general meeting, J. G. Lambton, M.P., in the chair; and among other resolutions was the following:—

"That this meeting is decidedly of opinion, from the evidence produced in various publications, by Mr. George Stephenson and his friends, as well as from the documents which have been read at this meeting, that Mr. Stephenson *did not* discover the fact that explosions of fire-damp will not pass through tubes and apertures of small dimensions; and that he *did not* apply that principle to the construction of a safety-lamp; and that the latest lamps made by Mr. Stephenson are evident imitations of those of Sir Humphry Davy; and that, even with that advantage, they are so imperfectly constructed as to be actually unsafe."

Strictly speaking, the discovery that flame will not pass through tubes of small dimensions is due neither to Mr. Stephenson nor to Sir Humphry Davy, though the latter did develop the immense importance to which the *principle* might be applied. The discovery was made by Dr. Tennant, assisted, it is believed, by Dr. Wollaston, (who however laid no claim to it,) while Sir Humphry was travelling on the Continent, but to him are due all the beneficial results to which it gave rise, and Mr. Tennant was too modest to assert his undoubted claim to the discovery of the principle.

Among the numerous gratifications of various kinds received by Sir Humphry, in presents and praises, from the Emperor of Russia down to the poor working colliers, " I never was more affected," says Davy, "than by a written address which I received from the working colliers, when I was in the North, thanking me, in behalf of themselves and their families, for the preservation of their lives."

Sir Humphry's acquirements were now to be exercised in a new line, though still in chemical experiments. Trials had been made, but without success, to unroll the ancient *papyri* deposited in the Museum of Naples. He had convinced himself, from products

obtained by distillation, that the nature of these manuscripts had been generally misunderstood; that they
had not been carbonized by the operation of fire, but
were in a state analogous to peat or to Bovey coal, the
leaves having been cemented into one mass; and he
concluded that the substance formed between them
would become a subject of obvious chemical investigation and destruction, and be acted upon by chlorine
or iodine.

Encouraged by the Cabinet, and after receiving an
audience and eliciting the approbation of the Prince
Regent, Davy set out for Naples on the 26th of May,
1818. He was received at Naples, and every attention
and facility was given towards furthering the pursuit of
his objects. On examining, the different appearances
of the papyri were attributed by the persons who had
the care of them to the action of fire. The conclusion
at which he had arrived, while in England, was confirmed still more by a visit he made to the excavations
that remained open at Herculaneum. He gave a detailed account of the condition in which he found the
papyri, but beyond that his research was not successful.
His biographer observes—"It will readily be supposed
that a failure in an investigation from which he had
anticipated so much advantage was not sustained by a
person naturally quick and irritable without some
demonstrations of impatience and dissatisfaction." But
enough—" the conduct of the persons at the head of this
department in the Museum" will suffice to account for
those demonstrations.

Sir Humphry Davy was created a Baronet on the
20th of October, 1818, and returned to England in
1820, and on the 19th of June in that year Sir Joseph

Banks, who, notwithstanding his increasing infirmities, had continued to discharge the duties of President of the Royal Society, expired at his villa of Spring Grove.

I have just related Dr. Wollaston's positive refusal to be nominated for the Presidency. Whether Sir Humphry Davy was invited to put himself forward as a candidate, or voluntarily offered himself, seems doubtful. In a letter to his friend Mr. Poole, a few days after Sir Joseph's death, is the following passage:—"I feel it a duty I owe to the Society to offer myself; but if they do not feel that they want me (and the most active members I believe do), I shall not force myself upon them." In the same letter he says, "I feel that the President's chair, after Sir Joseph, will be no light matter; and unless there is a strong feeling in the majority of the body that I am the most proper person, I shall not sacrifice my tranquillity for what cannot add to my reputation, though it may increase my power of being useful."

Davy was elected almost without opposition: Lord Colchester, having been, in his absence, put up by a very few members, had 13 votes out of about 160. In announcing this result to Mr. Poole, he says, "I glory in being in the chair of the Royal Society; because I think it ought to be a reward of scientific labours, and not an appendage to rank or fortune; and because it will enable me to be useful in a higher degree in promoting the cause of science."

It was a question discussed by the friends of Davy, how far his elevation to the chair of the Royal Society was calculated to advance the cause of science, or to increase the lustre of his own fame. It is a question that involves considerations, not only as relating to the

character of the person, but also to the constitution
and objects of the Society. In looking at the present
state and progress of science, and at the names of those
from whom the papers published in the 'Philosophical
Transactions' mostly proceed, we should be apt to say
that the personal character of the President, his general
and intellectual acquirements, and his knowledge of
mankind, accompanied by an easy and courteous manner
of receiving and dealing with them, is more likely to
ensure high success in promoting the objects of the
Society than the mere man of science. Sir Joseph
boasted not of science, yet where, or when, shall we find
his equal as a President?

As President of the Royal Society, Sir Humphry
Davy did not make himself popular. He was punctual,
however, in the fulfilment of the duties of his new office,
constant in his attendance at the meetings of the Society,
and dignified in his conduct and deportment when in
the chair; in council he took a prominent part, proposed
several resolutions, supported them with firmness, and
decided impartially on plans and propositions brought
forward by the members. It was not to be expected,
however, considering the claims on the time and
attention of a President of the Royal Society, and
the private calls made upon him, that he could, if so
disposed, prosecute his experiments and discoveries in
chemical subjects, as he otherwise might have done.
Dr. Wollaston was a man who felt this difficulty, and
no doubt was chiefly, if not wholly, governed by it in
his refusal to accept the chair of the Royal Society.

These two eminent men were of temperaments widely
dissimilar in every respect, so that even in their studies
they pursued different paths. Dr. Paris says, "It has

been observed that the chemical manipulations of Wollaston and Davy offered a singular contrast to each other; but in contrasting the genius of Wollaston with that of Davy, let me not be supposed to invite a comparison to the disparagement of either, but rather to the glory of both; for by mutual reflection each will glow the brighter. Davy was ever imagining something greater than he knew; Wollaston always knew something more than he acknowledged. In Wollaston the predominant principle was to avoid error; in Davy it was to discover truth. The tendency of Davy on all occasions was, to raise probabilities into facts; while Wollaston continually made them subservient to the expression of doubt."

In 1823 the Navy Board made application to Sir Humphry Davy to procure the talents of the most scientific men in the kingdom to examine into the defect in the present mode of manufacturing copper-sheets for the purpose of sheathing the ships' bottoms, in order to secure to the Navy copper of the most durable quality, and such as will preserve the smoothest surface. Davy called for a committee, consisting of himself, Brande, Hatchett, Herschel, and Wollaston; all these philosophers had more or less been engaged in galvanic experiments, and therefore it was well known to them that by attaching pieces of tin, zinc, lead, or iron to copper, that metal will become electro-positive, and thus change its nature. Experiments proved to them that a piece of zinc attached to a sheet of copper will prevent corrosion of the latter in salt water, which was the subject of complaint.

In order to demonstrate the fact to the public, Sir Humphry requested that three models of small ships

might be made, and exhibited in the specimen hall of
the Navy Office in Somerset House; the copper on
one to be protected by bands of zinc, the second by
plates of wrought-iron soldered on the sheathing, and
the third to have the copper exposed without any kind
of protection.

These models were floated in large troughs filled
with sea-water, remained for many months, and were
examined, from time to time, by persons of the highest
scientific character, as well as by Naval Officers; and
so conclusive were the experiments, that the plan of
protection to the copper was forthwith carried out into
extensive practice, in ships both public and private.

It turned out, however, a decided and immediate
failure.   Dr. Paris, who is occasionally a little waggish,
says, " At length, paradoxical as it may appear, the
truth of Davy's theory was completely established by
the failure of his remedy—the copper was protected
from corrosion, but the bottoms of the protected ships
became extremely foul." The negative state of electricity caused the deposition of calcareous matter on the
copper: this encouraged the growth of sea-weed, and
the clustering of small marine shell-fish and insects.

When the protected ships were taken into dock,
their bottoms exhibited one mass of incrustations; and
when these were removed, the sheets of copper were
found to have sustained little or no loss of weight, veri-
fying in this respect the theory of Sir Humphry Davy,
but negativing its practical applicability for the intended
object. Nothing therefore remained but to restore the
protected ships to their former state; and in July, 1825,
a general order was given to that effect, to the great
mortification of Sir Humphry, who had taken to himself

the labouring oar, and had therefore the further vexation of a few squibs launched at him on the occasion.
It was stated by those who best knew him that the
failure exerted a marked, and unfavourable, influence
on his character.

Wollaston, who always preferred practical experiment
to theoretical reasoning, told me that, after several trials
on a small scale, and after much inquiry and attention
regarding the *poling*, or stirring up, of the melted
copper, at the dock-yard, he had come to the conclusion, almost to a certainty, that it was not sufficiently
poled, in order to get rid of impurities, and to render
the sheets more solid and durable.   It was put in practice, and I believe succeeded, as the complaints of corrosion were less frequent than heretofore.

In 1825 Sir Humphry was seized with an attack of
apoplexy, which left a great weakness upon him, yet
did not prevent him from officiating in the chair, at the
anniversary meeting of the 30th of November, 1826,
but it was for the last time; after which he was unable
to prosecute any scientific labours.

At the next anniversary meeting, 30th of November,
1827, a royal medal was awarded to Sir Humphry
Davy, who was then on the Continent, when Mr.
Davies Gilbert delivered an address, embracing the
main points of Sir Humphry's discoveries, and thus
concludes:—

" Sir Humphry Davy having last year communicated
a paper to the Society in continuation of his former
inductions and generalization on chemical and electrical
energies, there cannot be a doubt but that the only
obstacle against his then receiving a royal medal, on
the first occasion that the Society had it to bestow, was

H

his occupying this Chair.   That obstacle, unhappily for science, no longer exists; and the Royal Society takes this earliest opportunity of testifying their high estimation of those talents and of those labours which all Europe admires. We trust and hope, although our late President has been induced by medical advice to retire from the agitation of active public stations, that his most valuable life will be long spared ; and that energies of mind may still be displayed to this Society, and to the civilized world, equal to those which have heretofore rendered immortal the name of Davy."

Sir Humphry, however, never received his medal, owing to delays in the execution of the dies, which I shall explain, under the presidency of the Duke of Sussex, who removed the difficulty.

In the course of the last year he had amused himself by travelling about, fishing, shooting, and reading. In that year also, and in the preceding, he wrote a tract on fly-fishing, called ' Salmonia,' and another to which he gave the title of ' The last Days of a Philosopher,' both replete with sketches of natural history and philosophy, interspersed with poetical descriptions, evidently in imitation of the inimitable Isaac Walton, but not with the simplicity of that prince of fishers.

On the Continent he was again alarmed by symptoms of palsy, which were renewed in February, 1829, by a stroke that had nearly proved fatal, being then at Rome, on a continental travel, accompanied by Dr. Tobin, who thought it right to apprize Dr. Davy at Malta, and Lady Davy in London, both of whom arrived in Rome in the month of March ; and on the 30th of April Sir Humphry, accompanied by his wife and brother, left Rome, and on the 29th of May reached Geneva. By this time he had gained the power over his limbs, and was able to lie upon the sofa. "It seems impossible for him," says Dr. Tobin, "to exist without being read to, and in one day I read Shakespeare to him for *nine* hours."

At Geneva on the 29th of May the following passage is recorded in Dr. Tobin's diary :—" I quitted Sir Humphry yesterday evening, after having read to him as usual till about ten o'clock. Our book was Smollett's *Humphrey Clinker*, and little did I think it was the last book he ever would listen to. He seemed in tolerable spirits, but on going to bed was seized with spasms, which however were not violent, and soon ceased. I left him when in bed, and, bidding me 'good night,' he said I should see him better in the morning. Lady and Dr. Davy also quitted him. At six o'clock this morning Lady Davy's man-servant came to my room, and told me that Sir Humphry was no more. I went down to his room instantly, and found that his words were, alas! but too true."

A memorandum relating to the last hours of his brother is given by Dr. Davy :—" On the 28th of May we arrived at Geneva; at five he dined at table; after dinner he was read to; at nine he proposed to go to bed. In undressing he struck his elbow against the projecting arm of the sofa on which he sat. The effect was very extraordinary; he was suddenly seized with an universal tremor; he experienced intense pain in the part struck, and a sensation, he said, as if he were dying; he was got into bed as soon as possible. The painful sensations quickly subsided, and in a few minutes were entirely gone. There was no

mark of hurt at the elbow, no pain or remaining
tenderness; a slight feverish feeling followed, which
he thought little of; he took an anodyne draught of
acetate of morphia, and then desired to be read to,
that he might be composed to sleep by agreeable
images.

"About half-past nine he wished to be left alone, and
I took my leave of him for the night, and for ever on
earth.   His servant called me about half-past two,
saying he was taken very ill.   I went to him immediately : he was then in a state of insensibility, his respiration extremely slow and convulsive, and the pulse
imperceptible ; he was dying, and in a few minutes he
expired.   In death his countenance was composed, and
of its mildest expression, indicative of no pain or suffering.   This fatal moment was about three A.M. on the
29th of May."

Mr. Brande, in his 'History of Chemistry,' after
describing the labours and the great power of mind
that distinguished Davy and Wollaston, observes—

" The loss of two such men as Wollaston and Davy,
within so short a period, and scarcely beyond the prime
of life, was a serious national calamity ; the sketch I
have given of their labours is a most imperfect outline,
but it would have been improper and indecorous to
have dismissed this brief ' History of Chemistry ' without such notice."

The following are a few of his Papers, out of more than forty in the ' Transactions,' from 1821 to 1830 inclusive.

An Account of some Galvanic Combinations formed by Single Metallic Plates and Fluids, analogous to the Galvanic Apparatus of M. Volta.

On the Constituent Parts of Astringent Vegetables, and on their Operation in Tanning.

The Bakerian Lecture: On some Chemical Agencies of Electricity.

The Bakerian Lecture: New Analytical Researches in Chemistry, with Observations on Chemical Theory.

On Muriatic Acid, with Experiments on Sulphur and Phosphorus, made in the Laboratory of the Royal Institution.

On a Combination of Oxymuriatic and Oxygen Gas.

On a New Detonating Compound.

Some Experiments on a New Substance, which becomes a Violet-coloured Gas by Heat.

Further Experiments and Observations on Iodine.

Some Experiments on a solid Compound of Iodine and Oxygen, and on its Chemical Agencies.

On the Fire-damp of Coal-Mines, and on Methods of Lighting the Mines so as to prevent Explosion.

An Account of an Invention for giving Light in Explosive Mixtures of Fire-damp in Coal-Mines by consuming the Fire-damp.

Further Experiments on the Combustion of Explosive Mixtures confined by Wire-gauze, with some Observations on Flame.

On the Fallacy of the Experiments in which Water is said to have been formed by the Decomposition of Chlorine.

Section V.

Mr. Davies Gilbert.

The name of the family to which Mr. Davies Gilbert paternally belonged was Giddy, which he retained for many years. He was an only child, and was born at St. Erth, in March, 1767. Was educated partly at a grammar-school in Penzance, but mainly, by the care and attention of his father, under the paternal roof. He thence proceeded to Oxford, and was admitted as a gentleman commoner of Pembroke College, 12th of April, 1785; took the degree of Master of Arts in 1789. Such was the proficiency made by Mr. Giddy while at Oxford, that Dr. Parr, in writing to the late Master of Pembroke, speaks of him, then twenty-three years of age, as " the Cornish Philosopher;" and adds, " he deserves that name." Indeed, while at Oxford, he was most assiduous in his various studies. He regularly attended lectures on anatomy, chemistry, mineralogy, botany, geometry, and astronomy; and devoted himself with great diligence to the study of mathematics and the abstract sciences. It was said he made it his boast, with becoming pride, that he was the first student of his class, in the University of Oxford, who had ever read the *Principia* of Newton.

At Oxford he contracted an intimacy with Dr. Thomas Beddoes, notorious for his violent democratical principles, which Giddy always abhorred. Beddoes, however, had, as I have before stated, through Giddy's recommendation, the merit of bringing into public notice the talents of Sir Humphry Davy.

On leaving Oxford Mr. Giddy visited the metropolis, made acquaintance with some of the most learned and scientific men, and in 1791 was elected a Fellow of the Royal Society, and also of the Linnæan Society. In 1792 he served the office of high-sheriff of Cornwall. In 1804 he was elected for the borough of Helston, and in 1806 was returned for Bodmin, and sat for that borough until December, 1832, twenty-six years. His steady and persuasive manner in conducting an argument or settling a dispute occasioned him references without end, but his patience was untired. A friend once had recommended him to a Quaker on important business. " Thou sent me," he says, " a Mr. Giddy: I assure thee, if Giddy be his name, he was anything but giddy in his business and character."

In the year 1808 he married Mary Anne, the only daughter and heiress of Thomas Gilbert, of Eastbourne, in Sussex, and Mr. Giddy now took the name and arms of Gilbert only. As a senator, Mr. Gilbert was considered one of the most assiduous that ever sat in the House of Commons, and was probably unequalled for his important services on committees. The numerous parliamentary investigations (more especially such as were connected with arts or science) in which he took a prominent part, form lasting memorials of his profound learning and indefatigable perseverance; and the application of his knowledge to practical purposes was attested by the active interest he took in most of our great national works—in the ancient usages and customs of the mines of Cornwall, the Plymouth Breakwater, and the Pevensey Level.

When the Presidency of the Royal Society became vacant on the death of Sir Joseph Banks, Mr. Davies Gilbert was one of those looked up to as well qualified for that office, but was finally selected to be Treasurer.

In the early part of 1827, when Sir Humphry Davy was obliged to quit England on account of ill health, Mr. Gilbert took the chair as Vice-President at nearly every meeting of that session. Sir Humphry's indisposition extending to the next session, he found it absolutely necessary to retire, and Mr. Davies Gilbert was then chosen President, to the great satisfaction of the body at large, and especially of the more scientific members.

Sound and extensive acquirements in every department of science, courtesy of manners, and kindness of disposition, rendered him every way fitting for the high situation to which he had been chosen. But after the long and liberal reign of Sir Joseph Banks, who might be said to have kept open house in town, where he was, at all times, accessible to all inquirers, and hospitable to all visitors, it could hardly be expected that a successor equal to him in that respect would be found. Mr. Gilbert possessed all the requisites, which science and literature could bestow, for filling the prominent station to which he had been called; but he had no large and hospitable mansion in the metropolis, for the reception of the numerous guests by whom he would have been surrounded; and the government, or even the management, of such a body as the Royal Society required a more rigid and commanding deportment than nature and his limited commerce with the world had bestowed on him.

In the course of Mr. Gilbert's presidency he received
a communication from the Treasury, announcing that it
was the intention of Government to appropriate a part
of the house in St. James's Park, built for the Duke of
York, to the use of the Royal Society. I was one of
the deputation appointed to view the apartments and
report on their eligibility. Mr. Gilbert took a private
view of them, but gave no opinion—that of the Committee was that they were wholly unsuitable, and that
there was not a single room in the house fit for the
meetings of the Royal Society.

A trust of greater importance devolved on Mr.
Gilbert, while in the chair of the Royal Society. By
the will of the Right Honourable and Reverend Francis
Henry Earl of Bridgewater, who died in February,
1829, it was directed that eight thousand pounds should
be invested in the funds; which sum, with the dividends
accruing thereon, was to be at the disposal of the
President, for the time being, of the Royal Society of
London, to be paid to the person or persons nominated
by him, which person or persons should write, print,
and publish one thousand copies of a work ' On the
Power, Wisdom, and Goodness of God, as manifested
in the Creation ;' illustrating such work by all reasonable
arguments ; as, for instance, on the variety and formation of God's creatures in the animal, vegetable, and
mineral kingdoms; the effect of digestion, and thereby
conversion ; the construction of the *hand* of man ; and
on an infinite number of other subjects : as also by
discoveries, ancient and modern, in arts, sciences, and
the whole extent of literature.

It need scarcely be said that Mr. Gilbert felt strongly
the great reponsibility attaching to this trust, and, very

wisely, at once decided on calling in and requesting the assistance of the Archbishop of Canterbury and the Bishop of London, in order to determine the best mode of carrying into effect the intentions of the testator. Acting under their advice, Mr. Gilbert appointed the following eight gentlemen to write each a separate treatise on the subjects which follow :—

1. *The Rev. W. Whewell, F.R.S.*—Astronomy and General Physics, considered with reference to Natural Theology.

2. *The Rev. Thomas Chalmers, D.D.*—On the Adaptation of external Nature to the Moral and Intellectual Constitution of Man.

3. *John Kidd, M.D., F.R.S.*—On the Adaptation of external Nature to the Physical Condition of Man.

4. *Sir Charles Bell, F.R.S.*— The Hand: its Mechanism and vital Endowments, as evincing Design.

5. *Peter Mark Roget, M.D., F.R.S.*—On Animal and Vegetable Physiology.

6. *The Rev. Dr. Kirby, F.R.S.*—On the History, Habits, and Instincts of Animals.

7. *William Prout, M.D., F.R.S.*—On Chemistry, Meteorology, and the Function of Digestion.

8. *The Rev. William Buckland, F.R.S.*—On Geology and Mineralogy.

Notwithstanding the advice of the two dignified prelates here mentioned, and the good sense and caution of the President, Mr. Gilbert, the plan adopted failed to meet anything like general approbation. Some said that the intentions of Lord Bridgewater had been misrepresented; that he had intended one work only

SECT. V.]   MR. DAVIES GILBERT.   107

to be written, and not eight or any other number; and that, if no one person was to be found competent to execute the whole design, two or more might be called in to assist. Other objections were also made, but the most serious and well-founded was that the works were published in so costly a shape as to be public only to the rich; and never, I believe, were the intentions of a pious and benevolent testator more completely frustrated.

Mr. Gilbert possessed a good memory and a great variety of knowledge, and was therefore a most agreeable companion. It was said of him, "That his most endearing talent was his power of conversation. It was not brilliant: it was something infinitely beyond, and better than, mere display: it was a continued stream of learning and philosophy, adapted with excellent taste to the capacity of his auditory, and enlivened with anecdotes to which the most listless could not but listen and learn."   Dr. Buckland, the present Dean of Westminster, said of him at the Geological Society, "His manners were most unaffected, childlike, gentle, and natural. As a friend he was kind, considerate, forbearing, patient, and generous; and when the grave was closed over him, not one, man, woman, or child, who had been honoured with his acquaintance, but felt that he had a friend less in the world: enemies he can have left not a single one." It would hardly be supposed that a man of so mild and courteous a character, so correct in all his dealings public and private, and so generally beloved, should have found himself opposed and thwarted by an active section of his colleagues (not numerous, I think), but certainly so it was. His excessive courtesy and desire

to please everybody had, as in the long run generally
happens, the effect of displeasing a portion of those
with whom he had to deal.

I have heard that when Mr. Gilbert was elected to
the Chair he told his most intimate friends that he had
made up his mind not to extend his presidency beyond
the period of three years; and at the commencement
of the third year he announced his intention of retiring at the next anniversary of the Society, on the
30th of November, 1830, and he did so. This may
have been the result of a previous resolution, but even if
it were not, circumstances had arisen that rendered his
continuance in the chair very inconvenient. He had
been, in the preceding year, apprehensive of having the
mortification to find the President's recommendation for
the choice of the Council and officers, for the first time,
disregarded. There was, I believe, no such danger;
but the very idea of any kind of contest was enough to
influence him, and having reason to suppose the Duke
of Sussex had expressed a wish to be placed at the
head of British science, Mr. Gilbert gladly availed himself of so good an excuse for resigning the chair. In
consequence of his announced resignation and of these
differences, meetings of several of the most influential
Fellows of the Society were held, to arrange amicably
the choice of a new President and Council.

" It was moved by Mr. Herschel, and seconded by
Mr. Faraday, that the President and Council be recommended to take into their consideration the propriety
of making out a list of fifty Fellows, out of whom they
would advise the Council and officers for the ensuing
year to be chosen; and that such list should contain
—first, the names of the members of the existing Council,

stating whether there were any vacancies from death or
absence; secondly, the names of twenty-nine other
Fellows, out of whom they would advise the Society to
select persons to fill up the vacancies on the day of
election."

This resolution was acted on by the Council, who
also recommended persons for the offices of Treasurer
and Secretaries; but the selection of a President was
left in the hands of the Society at large. Several persons were thought of for this office, but as the day
of election approached two candidates only appeared.
These were the Duke of Sussex and Mr. (now Sir John)
Herschel. At the beginning of November the following
advertisement in favour of the latter appeared in the
public papers, signed by sixty-three Fellows:—

"The undersigned Fellows of the Royal Society,
being of opinion that Mr. Herschel, by his varied and
profound knowledge and high personal character, is
eminently qualified to fill the office of President, and
that his appointment to the chair of the Society would
be peculiarly acceptable to men of science in this and
foreign countries, intend to put him in nomination on
the ensuing day of election."

Considerable exertion was of course used in favour
of the respective candidates; and the result seemed
doubtful.

On St. Andrew's day the attendance of Fellows was
unprecedentedly numerous, and included the principal
philosophers of the day, as also several individuals of
high rank and distinction. Mr. Gilbert took the chair
at twelve o'clock, and, after delivering his address, a
ballot was taken for the Council. The election fell

upon the Duke of Sussex, Sir A. Cooper, Col. Fitz-
Clarence, Captain Kater, Lord Melville, Sir G. Murray,
Sir Robert Peel, Dr. Roget, and Messrs. Barlow, Bar-
row, Cavendish, Children, Ellis, Faraday, Gilbert,
Lubbock, Peacock, Philip, Pond, Rennie, and Vigors.
A ballot for a President then took place: the result
was—

For the Duke of Sussex . . . 119
For Sir John Herschel 111

Mr. (now Sir John) Lubbock was chosen Treasurer,
and Dr. Roget and Mr. Children Secretaries. It
was resolved that a deputation should wait on his
Royal Highness to communicate the result of the election; after which Mr. Gilbert thanked the Society for
the attention shown to him during his Presidency, and
then vacated the chair. This election, which fell upon
gentlemen most, if not all, of whom had been in Mr.
Gilbert's former councils, showed, I think, that he had
had no real cause for alarm on account of the Presidential authority, if firmly exercised.

In 1839 his health had become much impaired; his
spirits also were evidently failing; yet he made a journey into Cornwall, where he presided for the last time
at the anniversary of the Royal Geological Society of
that county, of which he had been, in 1814, the founder,
and was President until his death. In the same year
he took leave of Oxford, after which he took a final
leave of London, retired to his house at Eastbourne in
November, 1839, where he died in the midst of his
family on the 24th of the following December.

Mr. Gilbert omitted no opportunity of being useful

to society, and more particularly to the inhabitants of his native county, Cornwall. He afforded great encouragement, by his pen and his purse, to the improvements in working the valuable mines in that county. He gave an important contribution to Cornish topography, by a corrected and improved edition of a previous but imperfect History by Hals, with additions from the MSS. of Tonkins and Whitaker. Mr. Gilbert added something of interest to every parish in the county, and introduced many valuable remarks on the family, history, and biography of the most eminent Cornishmen, ancient and contemporary.

As memorials of a language of which but little has been preserved, and for amusement, he translated a few ancient trifles. One was 'A Collection of ancient Christmas Carols,' with the tunes to which they were formerly sung in the West; printed by his own printing-press in his house at Eastbourne. Another piece was 'The Creation of the World, with Noah's Flood;' written in Cornish, in 1611.

Mr. Gilbert was a considerable contributor to the 'Philosophical Transactions.' Among others, a paper on the Catinary Curve, with tables for constructing the Menai Bridge; a second paper on Steam Engines, and a third on the Nature of Imaginary Curves. The Journal of the Royal Institution contains several of his papers, of considerable length, on the Vibration of Pendulums.

Section VI.

His Royal Highness the Duke of Sussex.

THE election of the Duke of Sussex and that of his successor have fully confirmed what I have always strenuously maintained, and have already observed— that it is not necessary the President of the Royal Society should be what is termed a philosopher, that is, eminently distinguished in any particular branch of science; nay, I doubt if it be desirable, as the mind dedicated to one object will naturally give a predomi- nance to that one, and an universalist in the whole range of arts and sciences would be still more objection- able. I myself should prefer a gentleman of literary and scientific tastes and habits, whose rank, manners, and fortune fit him for becoming the centre round which the learned and the ingenious of all ranks and conditions may most easily and unreservedly group themselves. In the selection of his Council such a President may combine the highest qualifications that, in each department of science, the country affords. Such were, as I have before said, the qualifications of Sir Joseph Banks; and, with all my respect for his succes- sors, I may venture to say — *quando ullum invenias parem.*

His Royal Highness was born at the Queen's palace, Buckingham House, on the 27th of July, 1773; was created a Knight of the Garter in 1786. Having pro- secuted his studies a few years at home, he was sent to the University of Göttingen, and thence proceeded to

travel in Italy. While there, and still under age, he
contracted a marriage with Lady Augusta Murray,
second daughter of the Earl of Dunmore, which was
solemnized at Rome on the 4th of April, 1793, and
afterwards in St. George's, Hanover Square, on the 5th
of December following. By this irregular marriage
(which was legally pronounced null and void in August,
1794), he displeased his Royal Father, and conse-
quently, as I suppose, adopted opposition politics. He
was a pretty constant attendant in the House of Lords,
voting and occasionally speaking with the Whigs, but
was best known to the public in the more useful and
popular character of patron of various public charities,
and as a very efficient chairman at their anniversary
dinners. He was elected, in 1816, President of the
Society of Arts, in which he always took great interest
up to his last illness; and frequently appeared before
the public at the distribution of prizes. On such occasions his addresses were all that could be desired; neat,
varied, forcible, and appropriate. In 1818 he received
the honorary degree of LL.D. from the University of
Cambridge, as a member of Trinity College.

All these circumstances had concurred in alleviating,
in some degree, the obvious objections to placing a
Prince of the Blood in the chair of the Royal Society,
and he was, as I have stated, elected on the 30th of November, 1830, by the insignificant, or I should perhaps
rather say, very significant majority of 8. It was to be
expected that among such a minority there would be
found a number of persons discontented, and those
the weightiest in public opinion, who had no particular satisfaction at having a Prince thus preferred
to a man of science; but the affability and good

humour of the Duke, and the kindness of his manners, conciliated many adverse feelings, and left him
but a few avowed opponents.   Yet I believe that
he never altogether forgot the smallness of his majority.   It was hinted, indeed, that he soon discovered
this gratuitous office, honourable as it was, to be bur-
thensome to one who held it with a limited income,
and it was said that the insufficiency of his means
was the reason for his ultimately giving up the chair;
but his Royal Highness took eight years to make
the discovery; and I can hardly suppose such a motive.
He may have given an additional dinner now and then
to a small select party at Kensington Palace, and a
soirée or two in the season; but these small claims,
of the Royal Society, on his hospitality could have
occasioned him very little, if any, additional expense.

No one could speak or write, whether English or
French, better than the Duke of Sussex.   No one
could discuss a subject, or debate a point, with better
effect than his Royal Highness; and in such cases he
was neither deficient in force of language nor in argument; and in the heat of debate he preserved his temper,
and never lost his good humour.

It was by his Royal Highness's intervention that
the affair of the medals, to which I have before
slightly alluded, was brought to a satisfactory conclusion, and here I think it right to give the details of that
transaction.

In 1825, during the presidency of Sir Humphry
Davy, Sir Robert Peel had communicated to him
the King's intention to found two gold medals, of the
value of fifty guineas each, to be awarded as honorary

premiums, under the direction of the President and Council of the Royal Society, in such manner as should, by the excitement of competition among men of science, seem best calculated to promote the objects for which the Royal Society was instituted.

Thanks were of course returned; and in the following meeting it was resolved to propose for his Majesty's approbation, that the presentation of medals should not be limited to British subjects; and that two medals should be struck from the same die, one in gold and one in silver.

These resolutions being approved by his Majesty, directions were given for the preparation of two medals to be ready for awarding in 1826; and two men of science were selected by the President and Council to receive the first medals; but, unfortunately, no medals were forthcoming. The two eminent men selected were Mr. John Dalton, of Manchester, F.R.S., for his development of what is called the Atomic Theory, in chemistry; and James Ivory, for his papers on Astronomical Refraction, and other mathematical illustrations of important parts of astronomy.

Mr. Weld, who has given the history of the royal medals, observes that Sir Humphry Davy must have regretted that the medals were not ready for delivery. He did not however mention the fact in his address; and the reader of that address would naturally suppose that the gentlemen selected by the Council for the prizes had actually received them. But nothing in fact had been done till the matter was revived seven years later by the Duke of Sussex.

The exact cause of this delay was explained in the Royal President's address from the chair on the 30th of

November, 1833. "It must be," he says, "well known
to you, Gentlemen, that the royal medals were not
adjudged during the first two years that I presided
over the Royal Society; and as there exist many
circumstances connected with the original grant and
distribution of those medals, I trust I may be allowed
to enter into some details respecting them." His Royal
Highness then gives an account of their original foundation, and afterwards proceeds as follows :—

"Mr. Chantrey, to whom, in conjunction with Sir
Thomas Lawrence, was intrusted the selection of the
subject for the medal, furnished the cast for the medal-
lion of the head of his late Majesty, which was to form
the obverse of it, while Sir Thomas undertook to com-
pose the design for the reverse. Unfortunately, that
distinguished artist, either from over delicacy or over
anxiety to produce a work of art worthy of the object
for which it was intended, or from that spirit of pro-
crastination which was unhappily too common with
him, delayed its execution from year to year, and died
without leaving behind him even a sketch of his ideas
respecting it, though the character of such a design as
would be at once classical and appropriate to the purpose was the subject of frequent conversation, and even
of favourite speculation, with him. From these and
other causes, to which it is not necessary for me now to
advert, it arose that, at the demise of his late Majesty,
although the adjudication of ten medals had been form-
ally announced from the chair of the Royal Society,
not even the dies were forthcoming, much less the
medals, for distribution to the various distinguished persons, some of them foreigners, to whom they had been
awarded.

" It cannot be necessary for me to impress upon you, Gentlemen, that the non-completion of an engagement so solemnly entered into with the whole republic of the men of science would have brought discredit, not merely upon the Royal Society, but upon the personal honour of a monarch of this country whose name it is our special duty, as Fellows of the Royal Society, to hand down unsullied to posterity, as our munificent patron and benefactor ; and as no funds had been placed at the disposal of our treasurer, nor in the hands of any other ostensible person, to meet the very heavy expenses which must be incurred for cutting the dies and furnishing the medals already awarded, I felt it to be my duty, when I succeeded to this chair, to recommend to the Council the suspension of any further adjudgment of the medals, until I could have the opportunity of ascertaining the nature of the commands which had been issued, concerning them, by the sovereign, through his official advisers, or otherwise, and also of taking the pleasure of his present Majesty respecting their continuance in future, and the conditions to which they should be subject. These inquiries terminated in the most satisfactory manner. On a proper application to those who were intrusted with the ultimate arrangement of his late Majesty's affairs, prompt measures, as far as it lay in their power, were adopted for the immediate fulfilment of every pledge which it was conceived had been given to the Royal Society, and to the public at large, in the name of George IV.

" The dies for the medals upon the old foundation are now completed, and ready for distribution. They bear upon the one side the likeness of his late Majesty, while the reverse represents the celebrated statue of Sir

Isaac Newton which is placed in the chapel of Trinity
College, Cambridge, with such emblematical accom-
paniments as seemed best calculated to indicate the
magnificent objects of the researches and discoveries
of that great philosopher, whose peculiar connexion
with the Royal Society forms the most glorious circumstance of its annals.

"It will be my first duty, Gentlemen, to distribute
the ten royal medals which have been already
adjudged, during the lifetime of his late Majesty, to
philosophers who are amongst the most illustrious in
this country, or in Europe. They form a glorious commencement of a philosophical chivalry, under whose
banners the greatest amongst us might feel proud to be
enrolled ; and though it may appear presumptuous in
me to hope that a constant succession of associates can
be found, either at home or abroad, who shall be considered worthy of being ranked with those noble founders of this order, yet I am confident that the Council
of the Royal Society will feel an honourable pride in
maintaining the character of the body whose members
are to be instituted by their choice."

Thus the Royal Society owed to the zeal and business-
like exertions of the Duke of Sussex the termination of
a state of neglect and suspense which was really dis-
graceful, and which a person of a less placid disposition
than Mr. Gilbert would not have tolerated.

On the anniversaries of the Society, the President is
called upon to officiate more largely than usual. He
has, in the first place, to address the assembled members
on anything that may regard himself personally, or the
Society at large; to announce the medals that have been

distributed, to whom, and on what account, with a brief history of the merits of the recipient. Then follows the obituary of the preceding year; with a short sketch of the transactions, the experiments, the discoveries, and whatever tends to the development of the character and the labours of the deceased members in the cause of science. In all these points, and particularly in the obituary, his Royal Highness's addresses were given with great feeling and accuracy, and in neat and appropriate language.

I shall venture to give a few extracts from his two last speeches, which were heard with considerable interest. In 1836 his Royal Highness had adverted to the possibility of his being, from defect of sight and ill health, obliged to retire from the chair; but in 1837 he was so much better as to be enabled thus to address the Society:—

" Gentlemen,

" When I last had the honour of addressing you from this chair, I ventured to express a hope that the happy restoration of my sight, and the continued possession of health, would have enabled me to discharge, with becoming regularity, the duties of President of this Society, during those portions of the year in which I am generally resident in London; the fulfilment, however, of that hope was unhappily frustrated by a long and dangerous illness, which confined me for several months to my apartments, and from the effects of which I have hardly yet entirely recovered. I trust, Gentlemen, you will pardon me if I look forward with brighter hopes to the prospects of another year, and if I hesitate to regard the unhappy experience of that which is past

as a premonition of the fate which awaits me in those
which are to come; if such were my assurance or reasonable fear, I should acquiesce in the duty and propriety of at once retiring from this chair, and no longer
soliciting the renewal of an honour which I have enjoyed for so many years; but if it should be the pleasure of that good Providence, whose chastisement and
whose mercies I have so often before experienced, to
disable me from presiding over this Society in such a
manner as might be considered necessary for the protection and maintenance of its just interests and dignity,
I should bow with humble resignation to the expression
of his will, and resign to other hands the discharge of
those duties for which I should feel myself no longer
qualified.

" Since the last annual session of this Society we
have lost, Gentlemen, a most munificent patron and
benefactor, by the demise of our late most gracious Sovereign King William IV., of whom it is difficult for
me to speak in terms which do justice to my feelings.
He was, indeed, not less distinguished by the exalted
station which he filled than by the warmth and sincerity
of his affections, as a husband, a brother, and a friend;
by the undisguised frankness and truth of his character
as a man; and as a monarch, by his patriotic zeal to
increase the efficiency and secure the permanence of
the great institutions of his country, and to extend to all
classes of his subjects the blessings of peace and knowledge and the protection of just and equal laws.   I
would gladly enlarge, if the time or the occasion would
permit me to do so, upon these and many other virtues
in the character of one who was so closely connected
with me by the ties of relationship and of duty; but I

am quite sure that I should fail in the expression both
of your sentiments and my own, if I did not acknow-
ledge, in becoming terms of respect and gratitude, the
especial patronage and protection which he extended to
the Royal Society, by the renewed grant of the two
annual medals which had been instituted by his royal
brother and predecessor, and by the enactment of such
statutes for their distribution as appeared to him best
calculated to stimulate the exertions of philosophers,
and to associate for ever the results of their labours,
with the publication of the Transactions of the Royal
Society."

He then proceeded to give an account of the life of
Mr. Henry Thomas Colebrooke, who had died in the
preceding year. It is too long for insertion here, but
I must not omit the concluding paragraphs:—

"Mr. Colebrooke continued the steady pursuit of his
oriental and scientific studies until nearly the close of
his life, and even when the progress of his infirmities
confined him almost to his bed. He was one of the
founders of the Asiatic and Astronomical Societies, and
a short time before his death he gave to the library of
the India House his incomparable collection of Sanscrit
and Asiatic manuscripts, which had been collected at
an expense of nearly 10,000l., with the noble view of
preserving them for ever from the danger of dispersion
by the fluctuating accidents of inheritance.

"Mr. Colebrooke was probably, with one single exception, the greatest Sanscrit scholar of his age; and
when we take into account his great acquirements in
mathematics and philosophy, and in almost every branch
of literature, combined with the most accurate and
severe judgment, and also his great public services in

situations of the highest trust and responsibility, we shall
not hesitate to pronounce him one of the most illustrious
of that extraordinary succession of great men who have
adorned the annals of our Indian Empire, the deaths
of so many of whom it has been my misfortune to record in my recent addresses from this chair."

Another short obituary from the same speech must
suffice to exhibit the manner in which his Royal Highness was in the habit of commemorating our departed
colleagues :—

"Dr. John Latham reached the extraordinary age
of ninety-seven years, having enjoyed the full possession
of his faculties and almost unbroken health until within
a few days before his death; he was the father of the
Royal and Antiquarian Societies, and it is sixty-seven
years since his first paper, on a medical subject, was
published in our 'Transactions.' He was the author of
many papers on antiquarian subjects; but his favourite
study, throughout life, was natural history, and particularly ornithology. He published in 1781 his 'General Synopsis of Birds,' in six volumes quarto, and
afterwards two supplementary volumes. In 1792 he
published his 'Index Ornithologicus,' a complete system
of ornithology, arranged in classes, orders, genera, and
species, in two volumes quarto. At the age of eighty-two
he commenced his 'General History of Birds,' a magnificent work in eleven volumes quarto. He was a man
of very systematic habits and most amiable character,
the tranquil course of whose long life was neither disturbed by scientific or professional jealousies, nor
embittered by the want of those enjoyments which
competence and domestic happiness and virtue alone
can confer."

At the next following anniversary, the 30th of November, 1838, his Royal Highness announced his retirement, and took leave of the Society in a long and able
address.

He modestly alluded to the many occasions on
which he had been wanting in the duties of the office,
by the severe and long-continued visitations of disease
and infirmity, but felt the assurance that, by the alacrity
and zeal of the very able and efficient officers of the
Society, its interests would have suffered no detriment
by his absence.

He then noticed some of the most important proceedings of the Society during the last year; that
her Majesty had consented to become the patron of the
Society, had confirmed the continuance of the royal
medals which had been instituted by George IV., and
re-granted by William IV.; and had signed the Charter-
book at St. James's Palace in presence of the President and Council, who had the honour of kissing her
Majesty's hand.

He then adverted to the laudable example of the
Storthing or National Assembly of Norway, a body
composed partly of peasants, and representing one of the
poorest countries in Europe, undertaking the charge of
a magnetical expedition to Siberia, on the recommendation and under the direction of their distinguished countryman Hunsteen; and his Royal Highness added, evidently with more of approbation than the majority of
his audience was disposed to concur in, that they had
done so " at the same time that they refused a grant
of money to aid in building a palace for their sovereign."
Such a mode of putting science and loyalty into opposition with each other, seemed strange in the Royal

President of a Royal Society, which had just before been called upon to express its gratitude for royal patronage and royal distinctions.　He next adverted, more gracefully, to the return of Sir John Herschel to this country after an absence of several years, devoted, from a sense of filial duty, to the completion of that great task which he felt to have been transmitted to him, as an inheritance, by his venerable and illustrious father—and his Royal Highness concluded this part of his address by saying, " It was chiefly as an expression of the deference paid by the Government of this country to the opinions and wishes of the scientific world that I rejoiced in being authorized and requested by the Prime Minister of the Crown to offer to Sir John Herschel the rank of Baronet, on the occasion of the coronation of her Majesty, though well convinced that such an accession of social rank was not required to give dignity to one whose name is written in the imperishable records of the great system of the universe."

He next alluded to the probable successor to the chair which he was about to resign.　" It would ill become me," he said, " while gratefully acknowledging my sense of your past kindnesses towards myself, to venture to refer to the name of my presumed successor in the chair of this Society in any terms which might be interpreted as an undue anticipation of the result of this day's proceedings, or as appearing to interfere with the free use of the franchise which every Fellow possesses and is expected and required to exercise; but I cannot be ignorant of the various accomplishments, the courteous and unassuming manners, the warmth of heart, and active benevolence which distinguish the nobleman who has been nominated by the Council; and I rejoice most sincerely that the Society possesses amongst its members, as a candidate for your suffrages, one so well qualified to preside at your meetings, and to watch over your interests."

He concluded with enumerating the members deceased in the course of the year, and noticing their most distinguished labours; winding up his address, and at the same time closing his official labours, in the following sentence :—

"Gentlemen, I have now arrived at the last and most painful part of my duty in addressing you, which is most gratefully, and most respectfully, to bid you farewell."

Section VII.

The Marquis of Northampton.

[*The following are a few rough notes which were found thrown together by Sir John Barrow, who had expressed his intention of apprising Lord Northampton—for whom he entertained a very great regard—that he was preparing these sheets for the press, and of asking the Marquis for certain information necessary to enable him to give a brief memoir of his Lordship, as one of the six Presidents. This he had not an opportunity of doing : and these slight notices of a living person would not have been published without his Lordship's obliging consent to what seems indeed an almost indispensable chapter of the work.*]

Spencer-Joshua-Alwyne, Marquis and Earl of Northampton, was born 2nd January, 1790; was elected a Fellow of the Royal Society on 27th May, 1830 ; and President on the 30th November, 1838, on the resignation of the Duke of Sussex, and he first took the chair on the 2nd of December following.

After the transaction of some ordinary business, his Lordship took that earliest opportunity of announcing his intention of giving four soirées or evening parties during the session, to which he invited all the members of the Royal Society.

On the first anniversary meeting after his election

the new President had to announce in his address, and to explain, two important subjects. The first was that of the Antarctic expedition, then fitting out under Captain James Ross; the second was a system of simultaneous magnetical observations, to be carried on in our different colonies in every part of the world, on a plan suggested by the Baron Humboldt to the Duke of Sussex.

In order to check the indiscriminate admission of members, he also announced the establishment of new forms for the proposal of candidates; in these were to be stated the grounds on which the candidate was recommended.

Besides the four soirées that Lord Northampton gives during the season at his house in Piccadilly, to which he invites his private friends and foreigners, as well as members of the Royal Society; after the reading of the paper or papers at the weekly evening's meeting of the Society, he is in the habit of going with as many as please up to the Society's library, where they have tea and conversation. This is rather more than any of the recent Presidents have done towards keeping literary and scientific society together, but it is far from supplying the permanent and extensive encouragement given by such réunions as those of Sir Joseph Banks.

His Lordship's annual reports and speeches are well considered and well written. During the ten years that he has held the chair of the Royal Society he has given—at least I have heard nothing to the contrary—very general satisfaction; more than any President since Sir Joseph Banks. He is lively, cheerful, and agreeable in his manners, fond of hearing and not backward in telling pleasant stories, and when at the Club has something to say to every one. In short, he is just what Sir Joseph Banks would, after the formula of Dr. Johnson, have pronounced a *clubbable* man. A well-educated nobleman, a traveller over various countries of the Continent, one versed in the laws and institutions of his country, he is fit for any situation that an English gentleman could be qualified to hold, and that without more pretensions to science or philosophy than a good education and a taste for such studies would naturally justify.

The following extract from the address at the anniversary meeting of the 30th of November, 1846, may be given as a specimen of Lord Northampton's performance of almost the only public duty of the President :—

" The year that has just elapsed has been a very important one in the annals of science, both at home and abroad. On the Continent it has been remarkable for the discoveries of M. Schönbein, M. Le Verrier, and M. Gall; while the researches and calculations of M. Mädler, if confirmed and accredited by other astronomers, lead to results of such an extraordinary and gigantic character as to throw other discoveries into comparative shade.

" At home, we have the observations already made by Lord Rosse's unequalled telescope, the continuation of the bright line of research pursued by our illustrious Faraday, and the remarkable discovery of a younger chemist, Mr. Grove, who I trust has still a long course of scientific glory to run."

After noticing the splendid illustrated description of gigantic extinct mammalia in the British Museum, he proceeds to another topic:—

" We have seen that which has been little more than the instrument of amusement—an amusement which it is true is well worthy of a great naval people, I mean a yacht—converted into the means of adding largely to our knowledge of the marine zoology of the British seas. As President of the principal scientific society in England, I think it right to express the thanks of science to Mr. MacAndrew for the liberality of which Professor Edward Forbes has made such good use, and more especially for the example thus set to others.   This, Gentlemen, is an instance, among many that might be found, of the utility, for the purpose of extending science itself, of spreading, even among those who do not absolutely pursue it themselves, the feeling of its real interest and importance.   It is a proof that the rich and the powerful can, from time to time, advance knowledge by holding out a helping hand to its active cultivators.

*   *   *   *   *

" Gentlemen,—We must remember that, though we are a Royal Society, our true glory does not rest on our royal foundation, nor on royal patronage ; nor does it rest on the names of the illustrious nobles, of the eminent statesmen, or of those distinguished in art or literature who may have given lustre to the lists of our members ; nor yet does it arise from the array of foreign philosophers who have considered that it is a desirable reward of their discoveries to be honorary members of our body ; nor even does it rest on the

K

great lights of science, either still burning or extin-
guished in death, who belong or who have belonged to
our Society.   No, Gentlemen; our true glory must be
chiefly found in our scientific utility—in the manner
in which we have fulfilled our duties and promoted
the objects of our founders—and more especially must
we look for our true title of honour in our ' Transactions.'

" Gentlemen,—I am now arrived at the most agreeable part of my presidential duty—the pleasant task
of acknowledging and conferring honorary rewards on
scientific merit."   [Here, in awarding one of the medals
to Mr. Le Verrier, the discoverer of the new planet, he
expressed his regret that the distinguished foreigner
should not have been present; but rejoiced that he was
represented by the son of another illustrious foreigner,
Sir John Herschel, whom he thus addressed:—]

" Sir John Herschel,—I have great pleasure in committing to your charge this medal, which has been
awarded by the Council of the Royal Society to M.
Le Verrier.   It is well deserved by the genius that
foresaw the result, and the persevering calculations that
enabled M. Le Verrier to predict the exact quarter of
the heavens where a new planet must pursue its course,
in obedience to those general laws by which the
Almighty governs the universe.   There Mr. Gall's
telescope enabled him to verify the calculations of the
young French astronomer; and other observers have
since witnessed the existence of this new member of the
solar system.   Astronomy does not merely owe to M.
Le Verrier the knowledge of this new companion of
those planets which were known to man already, but it

also owes to him a bright confirmation of the truth of the Newtonian theory itself—a confirmation that must speak convincingly to the most sceptical and the most ignorant, if, indeed, in this case, there be any other scepticism than that of ignorance."

The President next presented three royal medals; *two* to Mr. Faraday, and *one* to Mr. Owen.

" Mr. Faraday,—It is an unusual honour that I have to announce to you to-day, and it is with unusual pleasure that I do so.

" The Council of the Royal Society has adjudged to you two medals at the same time for your late brilliant discoveries in the universal action of electricity and galvanism. If, however, the honour be unusual, such a long-continued sequence of scientific discovery, such a stream of electrical light, thrown in as it were on the dark places of science, by the genius and the persevering energy of one man, is still more singular. It is my agreeable duty to add that, in presenting you these medals, I consider that I do so to one to whom English science, and most especially the Royal Society, lies under the deepest obligations. The Royal Society was itself founded for the more extended and more accurate cultivation of natural knowledge ; and while it can boast in its ' Transactions ' of such papers as those for which it is indebted to you, its prosperity must be regarded as established on the surest basis.

" Mr. Owen,—It gives me great satisfaction to announce that the Council has awarded one of the royal medals to you for your very excellent paper on the Belemnite. It is a communication of the highest interest at the same time to the geologist and the palæontologist. It describes and explains the nature of an extinct animal, one portion of whose frame is found existing in different strata, while very slight indications of the remainder of its structure had been known to the world till a very recent period. It adds to our satisfaction, as an English Society, that the ample account of this animal given to us by your anatomical skill and experience is derived from remarkable specimens hitherto at least found in England alone. Their discovery has been owing to the formation of the Great Western Railway. In this instance, therefore, and probably in many future ones, this gigantic instrument, the child of modern engineering genius, is not only the means of rapid locomotion to the traveller, but also carries forward, with accelerated speed, the progress of physical science itself."

The notified intention of Lord Northampton to vacate the chair at the next anniversary has excited general regret, though we must admit that he has amply fulfilled his duty to the public by ten years' service in a gratuitous and not very thankful office, that imposes a certain restraint on his time and inclinations, and often subjects him to be thwarted in the views he might entertain for the benefit and respectability of the Society over which, and for the control of which, he is ostensibly placed as the head, yet is virtually controlled by what is called *his* Council.

———————

Lord Rosse, who pursues astronomical science with an energy, activity, and liberality worthy a President of the Royal Society, is mentioned as his successor, and in all personal respects there could be none better; but his Lordship's scientific head-quarters (if I may use the term), with his observatory, being fixed at his country-seat in Ireland, we can scarcely hope to find him so permanent and attractive a centre of the learned and scientific world, as if he resided, as all his predecessors have done, principally in the metropolis.

Section VIII.

Mr. Alexander Dalrymple, F.R.S.

Having thus given a slight account of all the Presidents of the Royal Society in my time, I shall now proceed to say something of some of my other colleagues in the Society Club.

Alexander Dalrymple, who was Hydrographer to the Admiralty in the earlier years of my Secretaryship, was the son of Sir James Dalrymple, Bart., of Hailes, and of Lady Christian, daughter of the Earl of Haddington, a very amiable and accomplished woman; and brother of the celebrated judge and historian Lord Hailes. Having left the school of Haddington before he was fourteen, and not being sent to any other, or to the University, Alexander's scholastic endowments were very limited. His eldest brother was wont to make him translate some of the Odes of Horace to keep up his Latin. But going abroad entirely his own master, before he was sixteen years of age, no progress was made in his Latin: he says, indeed, as he never found any use for it, he took no pains to keep it up.

His family had interest enough with Alderman Baker, Chairman of the East India Company, to obtain for him a writership in the Company's service, which suited young Dalrymple, who had conceived a strong desire to go to the East Indies, by reading 'Menhoff's Voyages' and a novel called 'Joe Thomson.' He was for a short time placed in an academy at Enfield to be instructed in

writing and accounts, qualifications respecting which
he was to be examined, and some demur was made to
this part of his certificate. He was, however, appointed
a writer on the Madras establishment. But here was
another defect. He wanted a few months of sixteen
years of age; and, when mentioned, Lady Christian
strongly objected to her son tacitly assenting to countenance what was untrue, though it was urged that the
spirit of the regulation was only to prevent infants being introduced as writers, and not to preclude a person
for the want of a few months. Alexander contented
himself with saying that this was the only instance of
his having been accessary to the cheating of the Company, if it could be so termed.

He arrived at Madras in May, 1753; was put under
the Storekeeper, and thence was removed to the Secretary's office; was noticed with great kindness by Lord
Pigot and by Mr. Orme the historian. While in the
Secretary's office, on examining the old records, he
found that the commerce of the Eastern Islands was an
object of great consideration with the Company; and
he became inspired with an earnest desire to obtain that
important object for England. So forcibly was he
struck with the idea of the advantages in a national
point of view, that he was induced to propose to
Commodore Wilson, commanding the Indian squadron,
and through him to Governor Pigot, to accede to Mr.
Dalrymple going in the 'Cuddalore' schooner to the
Eastern Islands on a voyage of general observation.

But as the secretaryship would become vacant the following year, Lord Pigot, considering that situation a more
desirable object, endeavoured to dissuade Dalrymple
from the voyage, though without success, as he remained

warm in the pursuit of a plan of national importance, as
he had long conceived it to be ; and considered the
voyage in contemplation as a new era in his life.

His course lay to the Sooloo Islands, beyond Borneo ;
he waited on the Sultan, and made a treaty with
him, and a contract with the principal persons for a
cargo to be sent on the East India Company's account,
which the natives agreed to receive at 100 per cent.
profit, and to provide a cargo for China, which they
engaged should yield an equivalent profit there.   He
returned to Madras in January, 1762, and the Governor
was so well pleased that he appropriated the ' London '
packet, for a fresh voyage, and appointed Dalrymple to
the command of her.   In his passage out he visited
Balambangan, an island to the northward of Borneo.
The small-pox he found raging among the Sooloos:
the disease was most destructive, the chief people were
dead, and Dalrymple decided on returning to Madras.
In his way he called at and took possession of the island
of Balambangan for the East India Company, which
had been granted by the Chief of the Sooloos.

Disappointed in his views of forming an establishment on the island, and in his plans for the commerce of
the Eastern seas, he contented himself for the time by
publishing a plan of his intended operations, being one
of nearly a hundred books and pamphlets, on a great
variety of subjects connected one way or another with
the Sooloo Islands and the South Seas; and with these
subjects he engaged the attention of Lord Shelburne,
who was then Secretary of State, and who expressed a
desire to employ him on these discoveries.   It is stated
in a Memoir of Dalrymple's services that, when it was
decided to send persons to observe the transit of Venus

in 1769, he was approved of by the Admiralty as a proper person to be employed on this service, as well as to prosecute discoveries in that quarter; but from some differences of opinion, partly owing to official etiquette respecting the employment of any person as commander of a vessel who was not a naval officer, and partly owing to Dalrymple's objections to a divided command, this design did not take place.

In writing to Dr. Hawkesworth, who had accused him of misrepresenting the Spanish and Dutch voyages in the South Seas to support his own ill-grounded conjectures, he says, "You have, indeed, passed over in silence whatever you thought could do me any credit. Notwithstanding the injury done me in depriving me of the command of the ship I had chosen for the voyage, on pretence that I had not been brought up in the Royal Navy, so far was I from refusing my assistance to those who were going, that I gave Mr. Banks all the information I could; and accordingly he carried with him the ' Account of Discoveries made in the South Pacific Ocean,' with the Chart, which I had printed several months before, though I did not publish it till after Bourgainville's return."

In the same year the Court of Directors appointed Mr. Dalrymple Chief of Balambangan and Commander of the ' Britannia,' and made him, moreover, a present of five thousand pounds. He does not appear, however, to have proceeded to his government, but remained at home making and publishing plans of the Eastern Seas and Sooloo Islands. Yet all his plans, and the labour he bestowed on them, were of no avail; and the commercial intercourse with the Eastern Islands, the nearest object of his heart, appears to have failed. His zeal, however, in these pursuits procured him the appointment of Hydrographer to the East India Company. He held that office many years, and when it was abolished, Dalrymple retired on a pension for life: but the multitude of tracts which he wrote must have furnished him with employment.

The fact, however, of his having held such an office brought him into notice for another of the same kind. It having been long in contemplation to have an Hydrographical Office at the Admiralty, this was at length established during the administration of Earl Spencer, and Mr. Dalrymple was appointed Hydrographer in 1795. His biographer tells us—"Little occurred in his history worthy of particular notice until the month of May, 1808; when, having refused to resign his place of Hydrographer to the Admiralty, on the ground of being no longer equal to its duties, and to accept of a pension, he was dismissed from his situation; and it is said that, in the opinion of his medical attendants, his death was occasioned by vexation arising from that event."

I wish the biographer had given us some more details of the thirteen years he thus slurs over. My official connexion with him as Secretary of the Admiralty only existed in the last four of them; and from our intercourse at the Admiralty and at the Royal Society Club, I cannot refuse my assent to the remark of his biographer, that Mr. Dalrymple had exhibited so many symptoms of decayed faculties, joined to an irritable habit, as to lessen the value of those services for which he had previously been so highly respected. He died June 19, 1808.

Mr. Dalrymple was in fact an impracticable and

obstinate man, and very difficult to be diverted from any plan or project he had conceived. Mr. Secretary Wellesley Pole, who was perhaps not particularly gifted with the *suaviter in modo*, used frequently to complain of his perverse temper; and I cannot think that Mr. Dalrymple was at all justified, at the age of seventy, to resent, as he did, the proposition of superannuation.

SECTION IX.

HENRY CAVENDISH, ESQUIRE, F.R.S.

MR. CAVENDISH, son of Lord Charles Cavendish, and grandson of William, second Duke of Devonshire, was born at Nice, on the 10th of October, 1731, where his mother, Lady Anne Grey, daughter of Henry, Duke of Kent, had gone for the recovery of her health.

Much as one must be desirous to learn the details of the early cultivation of the intellectual and moral character of one who became so eminently distinguished in the higher branches of science, in natural philosophy, in chemistry, in pneumatics, in mechanics, and astronomy, little appears to be known of the early part of Henry Cavendish's education. He was for some time at Newcombe's noted school at Hackney, and afterwards went to Cambridge; but is supposed to have acquired a taste for experimental investigation, in a great degree, from his father.

The early talents that he showed made his family desirous that he should take a part in public life, which, however, he refused to do, his mind having formed a strong bias towards the sciences, and more especially for mathematics, chemistry, and the principal branches of natural philosophy; that in which he most excelled was pneumatic chemistry, by which he succeeded in making the astounding discovery that water was not a simple element, but a composition of two airs; for, by burning oxygen gas with hydrogen gas in a glass

globe, by means of electricity, he found that the whole
original contents of the globe had disappeared, and that
water, equal in weight to the two gases taken together,
remained as the product of the combustion.

Cuvier, in his 'Eloges Historiques,' speaking of this
celebrated discovery, says,—

"Ainsi l'on peut dire que cette théorie nouvelle,
qui a produit dans les sciences une si grande révo-
lution, a dû sa première origine à une découverte
de M. Cavendish, et que c'est une seconde découverte du même savant qui lui a donné son dernier
complément."

Physical science was about this time rapidly pro-
gressing, and this grand discovery, in particular, with
some minor experiments, raised up both here and on
the continent a great number of rival competitors, employed in experimental chemistry, particularly in the examination of factitious airs, among whom were Priestley,
Watt, Black, and Lavoisier, but we are told that Lavoisier's attempt to intrude himself into the splendid discoveries of Cavendish and some other English chemists
was wholly unsuccessful; it seems to have had no effect
except that of throwing a slur on the reputation of the
French chemists, already sufficiently injured by a similar attempt of his to share in the discovery of oxygen.
"All men," says Lord Brougham, "held Cavendish's
name as enrolled among the greatest discoverers of
any age, and all lamented that he did not pursue his
brilliant career with more activity, so as to augment still
farther the debt of gratitude under which he had
laid the scientific world."

Yet he had done wonders. For fifty years nearly he
had continued to make discoveries, quietly and privately,

in various branches of physical science, contented to make them available to the public by having them printed in the *Philosophical Transactions.*

"Whatever the sciences revealed to Cavendish," Cuvier adds, "appeared always to exhibit something of the sublime and the marvellous. He weighed the earth—he rendered the air navigable—he deprived water of the quality of an element—and he denied to fire the character of a substance. The clearness of the evidence on which he established his discoveries, new and unexpected as they were, is still more astonishing than the facts themselves which he detected; and the works in which he has made them public are so many masterpieces of sagacity and of methodical reasoning, each perfect as a whole and in its parts, leaving nothing for any other hand to correct."

It is but just, however, to assign to Lavoisier the merit that is due to him. He also, it is admitted, rendered great service, by his labours, to chemical science and natural knowledge; he had a ready facility in reducing the knowledge of scattered and isolated facts to a system; but he is charged with being apt to seize upon the inventions of others. Lord Brougham takes care to enumerate what Lavoisier did *not* do. "He did not, like Black, discover carbonic acid, or latent heat—did not, like Priestley, discover oxygen—he did not, like Scheele, discover chlorine—nor, like Davy, the alkaline metals—nor, like Cavendish, by direct experiment, show how water and nitrous acid are constituted—nor, like Berthollet, explain of what ammonia consists." He has, however, left behind him a great name in France, and not much inferior to some of those above named.

Among all the discoveries of modern times that of Cavendish is the one that has made the greatest noise in the world. From the year 1781 to 1784 the names of the most eminent philosophers of England and France are mixed up in this question; but the original discoverers were, after much discussion and a few pretenders, reduced to two, of whom—Cavendish, or Watt—each had his advocates in England. The French invariably pronounced Cavendish, and no other, to be the sole discoverer.

The dispute in this country would appear to have arisen chiefly from some confusion of dates. The paper of Cavendish was read before the Royal Society in January, 1784. It was not till March, 1784, that Watt took any steps to advance his claim to the discovery; and it may reasonably be doubted whether he would have done so then, had not Cavendish's paper been read publicly more than two months before. It is not improbable that these two eminent men made the discovery independent of each other, and perhaps pretty nearly at the same time, and that the experiments of Dr. Priestley and some others on different airs may have given to both competitors the first hint.

Mr. T. Brande,* in his ' Manual of Chemistry,' in describing the origin and progress of chemical philosophy, observes on "the little influence of external circum-

---

* Mr Brande went regularly through the *Royal Institution* under the patronage of Sir Humphry Davy; was assistant in the laboratory; was then appointed lecturer; and in 1813 succeeded Sir Humphry as "Professor of Chemistry:" is now "Superintendent of Machinery and Clerk of the Irons" in the Royal Mint. Sir Humphry has pronounced Brande's *Manual of Chemistry* "the best collection and arrangement of chemical science existing."

stances upon the growth of inherent genius," and illustrates the position by comparing the two great contemporary luminaries of chemical science—Cavendish and Scheele. "The former was a leading person in the scientific circles of London, of noble family, and princely affluence; the latter, of humble origin, and with limited means, made up for the deficiencies of place and fortune by zeal and economy, and in the retirement of a Swedish village raised a reputation that soon extended itself over Europe."

Mr. Cavendish was a man of silent and extremely reserved habits. He appeared to have an imperfection in his speech, a difficulty in bringing out his words, which he uttered with a shrill disagreeable voice, of which he appeared to be aware, and on that account averse from conversation. To the same reason might be ascribed his extreme shyness. He seemed to me, and all his habits confirmed it, to consider himself as a solitary being in the world, and to feel himself unfit for society.

Yet here we have a man possessed of an ample fortune, an uncle having left him above a million sterling, in addition to his large paternal property, who adhered closely to his original line of study, still uninfluenced by the ordinary ambition of becoming a distinguished statesman, by a taste for luxuries and expensive habits, or indulgence in any kind of extravagance. He saw little or no company; and it appears indeed that, after his accession to so much wealth, he made no change in his daily pursuits or habits of life.

Even in private life and among friends he was unassuming, bashful, and reserved; he was peevishly impatient of the inconveniences of eminence; he detested flattery, and was uneasy under merited praise; he therefore shunned general society, and was only familiar in a very limited circle of friends. He was in the habit of dining almost every Thursday at the Royal Society Club, and there he always bore his faculties meekly; but his conversation, when he became interested, was varied and instructive; upon all subjects of science he was at once profound and luminous, and in discussion wonderfully acute.

It has been remarked that he was not fond of female society, and we saw reason at the Club to suspect that he rather despised those who had the weakness to be so.

In the neighbourhood of Clapham, where his residence was, his extreme shyness and reserve were the subject of remark, and two young ladies determined to put it to the test. He was in the habit of walking to his house by a footpath along the hedge of a field. These ladies watched the opportunity of meeting him in this path to see if he would notice them, or if they could draw him into conversation. Long before he approached them, they observed him to turn off the path at right angles, and to cross the newly ploughed field, wet and dirty as it was.

This extraordinary shyness, which is supposed to have prevented him from joining private society, can scarcely be deemed natural, because it had no operation where objects of science or literature were in discussion. Thus he very constantly showed himself on Sunday evenings at Sir Joseph Banks's and attended regularly the Royal Society Club, which appeared to afford him much pleasure, where, however, he rarely joined in general conversation. I once mentioned to Sir Joseph that a friend of his and of mine, Mr. Alexander, the excellent

L

draughtsman to the China embassy, wished to take a full-length portrait of Mr. Cavendish. "Could I ask him this favour?" Sir Joseph's answer was not encouraging. "You may ask him, and get a blunt refusal; I cannot, having had a refusal on a similar occasion. But wait a little, and I will contrive to bring you together; and you may then try Mr. Cavendish himself."

A few evenings after this, I sat at table opposite to Mr. Cavendish: during dinner he continued, as usual, perfectly silent; but as soon as the things were removed and the waiter gone, he said, "I understand, Mr. Barrow, you are a native of the neighbourhood of Cartmel?" "I was born, Sir, in Ulverstone, over the sands, and know Cartmel very well, and Holker Hall, with its beautiful grounds and gardens, which belongs, I believe, to Lord George Cavendish." "It did belong to him, Sir; but he left it to my father, from whom it descended to me, and will next go to another Lord George" (afterwards created Earl of Burlington). I said it was nearly forty years since I was in that country. He mentioned other property of the Cavendishes with which I was well acquainted.

I soon found that I had touched a string that would vibrate. A few evenings after this, he asked me if I knew Mr. Wilkinson, the great iron-founder, and the great acquisition of fine meadow-land that he had contrived to gain from Morecambe Bay? I said, "I have frequently seen and admired it;" observing, it was on the opposite side of the bay, and nearly facing Holker. "Do you think," said he, somewhat eagerly, "that I could succeed, on my side of the bay, to get as much land out of it as he has done?" I stated the difference of the two shores, that of Holker being precipitous and the water deep, whereas on the Furness side, where Wilkinson carried on his operations, the water was shallow and ebbed out a great distance from the shore. It appeared to me then that the only way to gain land from the bay on that side would be, if practicable, to divert the Crake river from Morecambe Bay into the Holker estate, and carry it into the sea below Flock-borough.

The old gentleman chuckled with his shrill laughter, and said it appeared to him worth a trial, and that when he journeyed to the north he should look at it; but he died before he set out on his journey, and I believe that Morecambe Bay and the shore of Holker have not been disturbed—unless, as I have understood, for great improvements made by the Earl of Burlington.

I could not, however, find a favourable occasion for proposing the portrait, and at last Alexander, who was bent upon having it, said, if I would invite him to a club dinner, he could easily succeed, by taking his seat near the end of the table, from whence he could sketch the peculiar great-coat of a greyish green colour, and the remarkable three-cornered hat, invariably worn by Cavendish; and obtain, unobserved, such an outline of the face as, when inserted between the hat and coat, would make, he was quite sure, a full-length portrait that no one could mistake. It was so contrived, and every one who saw it recognised it at once. I think Alexander told me he should leave it at the British Museum; but whether it be there I know not.

With the exception of his visits to the Royal Society Club, to the President's house, and to his own house at the extremity of Gower Street, facing Bedford

Square, I believe he seldom left his villa at Clapham. The sole furniture of his town house might be said to consist of a library, of which he was liberal enough in lending out books to those with whom he was but slightly acquainted, or to others who could produce a recommendation, but every volume so lent was entered in a book; and so regular was he in this respect, that whatever volumes he might himself borrow, to take down to Clapham, he was careful to enter into the loan-book. It does not appear that he received any visitors in his domicile at Clapham. An old man-servant and two or three women, with a coachman and footman, seem to have comprised his whole establishment. How, it may be asked, did he contrive to get over the day? The only answer is, that his intense and abstruse studies in physics, metaphysics, mathematics, chemistry, and, it may be said, the whole circle of the sciences, gave him ample employment, in addition to which he amused himself with the mechanical operation of making many of his own instruments.

To form some idea of his philosophical labours we need only turn over the leaves of the ' Philosophical Transactions,' where will be found at least twenty papers on as many different subjects, written at various times through a period of more than forty years—papers that have been described as " the most interesting and important that have ever appeared in that collection, expressed in language which affords a model of concise simplicity and unaffected modesty, and exhibiting a precision of experimental demonstration commensurate with the judicious selection of the methods of research, and with the accuracy of the argumentative induction."

Three of these papers contain experiments on fictitious airs, and several others on airs, but those two which excited the greatest attention at home and on the continent are inserted in the 'Philosophical Transactions' of 1784. "This paper," says Dr. Young, "contains an account of two of the greatest discoveries in chemistry that have ever yet been made public—the composition of water, and that of the nitric acid. The author first establishes the radical difference of hydrogen from nitrogen or azote; he then proceeds to relate his experiments on the combustion of hydrogen with oxygen, which had partly been suggested by a cursory observation of Mr. Warlsire, a lecturer on natural philosophy, and which prove that pure water is the result of the process, provided that no nitrogen be present. These experiments were first made in 1781, and were then mentioned to Dr. Priestley; and when they were first communicated to Lavoisier, he found some difficulty in believing them to be accurate. The second series of experiments demonstrates, that when phlogisticated air, or nitrogen, is present in the process, some nitric acid is produced; and that this acid may be obtained from atmospheric air, by the repeated operation of the electrical spark."

The descriptive inventions and experiments printed in the 'Philosophical Transactions' exceed twenty: six or eight are on different kinds of air, independent of those (oxygen and hydrogen) which he discovered to be convertible into water; and independent of that great experiment, performed by instruments of his own contriving, for measuring the density of the earth, which he found to correspond very nearly with the result of Dr. Maskelyne's experiments on the mountain Schahallien.

The whole existence of Mr. Cavendish was absorbed

in scientific pursuits.   Abounding in wealth, he knew none of its enjoyments, nor even of what the world calls its comforts.   It purchased him no friends.   The very few who casually entered his only dwelling-house, on Clapham Common, describe its desolate appearance, and its scanty and mean-looking furniture.   Sir Everard Home was present at his death, and was surprised at what he saw.   The owner, however, appears, from the whole tenor of his life, to have been content, and to have desired nothing better.   The sketch which Lord Brougham has drawn of this remarkable man may partly explain his eccentric habits :—

" He lived retired from the world among his books and his instruments, never meddling with the affairs of active life; he passed his whole time in storing his mind with the knowledge imparted by former inquirers, and in extending its bounds.   Cultivating science for its own sake, he was slow to appear before the world as an author; had reached the middle age of life before he gave any work to the press; and though he reached the term of fourscore, never published a hundred pages. His methods of investigation were nearly as opposite as this diversity might lead us to expect; and in all the accidental circumstances of rank and wealth the same contrast is to be remarked : he was a duke's grandson; he possessed a princely fortune; his whole expenditure was on philosophical pursuits; his whole existence was in his laboratory or his library.   If such a life presents little variety, and few incidents to the vulgar observer, it is a matter of most interesting contemplation to all who set its just value upon the cultivation of science, who reckon its successful pursuit as the greatest privilege, the greatest glory of our nature."

"It was probably," says Cuvier, " either the reserve of his manners or the modest tone of his writings that procured him the uncommon distinction of never having his repose disturbed either by jealousy or by criticism. Like his great countryman Newton, whom he resembled in so many other respects, he died full of years and honours, beloved even by his rivals, respected by the age which he had enlightened, celebrated throughout the scientific world, and exhibiting to mankind a perfect model of what a man of science ought to be, and a splendid example of that success which is so eagerly sought, but so seldom obtained."

It may be deemed an honour to Sir Humphry Davy to have placed on record the following just and honourable tribute to the memory of one so distinguished in science; and, though so noble of birth, so unassuming, and even obscure in private life:—

"Of all the philosophers of the present age, Mr. Cavendish was the only one who combined, in the highest degree, a depth and extent of mathematical knowledge with delicacy and precision in the methods of experimental research. It may be said of him, what perhaps can hardly be said of any other person, that whatever he has done has been perfect at the moment of its production. His processes were all of a finished nature. Executed by the hand of a master, they required no correction; and though many of them were performed in the very infancy of chemical philosophy, yet their accuracy and their beauty have remained amidst the progress of discovery, and their merits have been illustrated by discussion, and exalted by time.

" In general the most common motives which induce men to study are the love of distinction, of
glory, or the desire of power; and we have no right to
object to motives of this kind; but it ought to be
mentioned, in estimating the character of Mr. Cavendish, that *his* grand stimulus to exertion was, evidently,
the love of truth and knowledge. Unambitious, unassuming, it was with difficulty he was persuaded to
bring forward his important discoveries. He disliked
notoriety; and he was, as it were, fearful of the voice
of fame. His labours are recorded with the greatest
dignity and simplicity, and in the fewest possible
words, without parade or apology; and it seemed as if
in publication he was performing, not what was a duty
to himself, but what was a duty to the public. His
life was devoted to science, and his social hours were
passed among a few friends, principally members of the
Royal Society. He was reserved to strangers; but
where he was familiar, his conversation was lively and
full of varied information. Upon all subjects of science
he was luminous and profound, and in discussion wonderfully acute. Even to the very last week of his life,
when he was nearly seventy-nine, he retained his activity
of body, and all his energy and sagacity of intellect. He
was warmly interested in all new subjects of science;
and several times in the course of last year witnessed,
or assisted in, some experiments which were carried on
in this theatre [the Royal Institution] or in the laboratory below.

" Since the death of Newton, if I may be permitted
to give an opinion, England has sustained no scientific
loss so great as that of Cavendish. Like his great pre-

decessor, he died full of years and glory. His name will be an object of more veneration in future ages than at the present moment. Though it was unknown in the busy scenes of life, or in the popular discussions of the day, it will remain illustrious in the annals of science, which are as imperishable as that nature to which they belong; and it will be an immortal honour to his House, to his age, and to his country."

Mr. Cavendish died at Clapham, on the 10th of March, 1810, in the eightieth year of his age, after a short illness, probably the first which he ever suffered. Lord Brougham observes that "his habit of curious observation continued to the end. He was desirous of marking the progress of disease, and the gradual extinction of the vital powers. With this view, that he might not be disturbed, he desired to be left alone. His servant, returning sooner than he had wished, was ordered again to leave the chamber of death, and when he came back a second time he found his master had expired."

This may be correct as to the main point, but the concluding details are not exactly so. Sir Everard Home, who was present at his death, gave me the following account. He said that Mr. Cavendish's servant came to his house in town, in great haste and alarm, and prayed him with all speed to hasten down to Clapham, as his master, he was quite sure, was dying, and had sent him out of the house, ordering him not to come near him till night, as he had something particular to engage his thoughts, and did not wish to be disturbed by any one. Sir Everard set off immediately for Clapham, anxious to ascertain the state in which his old friend might be, and he

ordered the servant to accompany him.   He found him
in bed, very much exhausted, and apparently in a
dying state.   Mr. Cavendish seemed rather surprised
to see him there; said that Sir Everard could be of no
use to him, for that he was in a dying state; and
blamed his servant for bringing him down from town,
for that at eighty years of age he thought that any
prolongation of life would only prolong its miseries.
Sir Everard insisted on remaining with him during the
night; and he ordered the female servant to prepare
and place a couch by his bedside.   The patient
remained very tranquil, and shortly after daybreak
departed this life.

When Sir Everard took a view in the morning of
this desolate and forlorn abode, the residence of one of
the most eminent men of science and philosophy that
this or any other country ever produced, overwhelmed
with riches which he never seemed to have enjoyed, it
occurred to him that some of the old chests of drawers,
cupboards, trunks, &c., might contain the valuable re-
lics of three wealthy generations, and that, until
notice of his death could be conveyed to some of the
Cavendish family, the three or four servants left in the
house might from curiosity, if not from other motives,
rummage these depositories.   Sir Everard therefore
desired the old servant who had returned with him to
bring him all the keys he could find, and to go round
with him, and open all that he could with the keys, in
order to ascertain any and what valuable property, or
papers of importance, they might contain, and to take
a note of them.   In one of the chests of drawers they
found many oldfashioned articles of old jewellery, parts

of embroidered dresses, &c., and, among other valuable
articles, an old lady's stomacher, so beset with diamonds
that, when it came to be examined and valued, I think
Sir Everard mentioned its worth at something like
twenty thousand pounds.   Sir Everard was anxious to
convey his report to the proper quarter—I think to
the late Earl of Burlington.

Section X.

Mr. Smithson Tennant, F.R.S.

Mr. Smithson Tennant, a distinguished chemist, born at Selby, in Yorkshire, on the 30th of November, 1761, the only child of the Rev. Calvert Tennant, who was a Fellow of St. John's College, Cambridge; determining to bring up this only son in the same line with himself, he set about teaching him Greek, when only five years of age.　Four years after this, however, the youth had the misfortune to lose his father, and, before he reached man's estate, his mother also, who, while riding out, was thrown from her horse and was killed on the spot—a fate which, as we shall see, was also that of her son.

Tennant had been sent to different country grammar-schools, and was reported to be indolent and indifferent even to boyish amusements: he however made progress in his education, having been fond, as a child, of read-ing books of science, especially such as had figures of explanation in them, and deriving great amusement from making experiments with such as were described in them; and when at school in Tadcaster, he took great delight in attending a course of lectures, by Mr. Walker, on experimental philosophy, that were given there.　His last school was at Dr. Croft's, in Beverley, where he entered little into the ordinary pursuits of his school-fellows, his habits being retired and unsocial.　Here, however, he derived great profit from an excellent library attached to the school, and among other books is said to have studied with great avidity Sir Isaac Newton's 'Treatise on Optics.'

It was obvious that he possessed an innate love of science, and that practical philosophy in particular was the main object of his desires. Among other branches of science, chemistry was the great attraction, and he made application to his friends to place him under the immediate instruction of Dr. Priestley, who was at this time enjoying deserved reputation for his experimental discoveries. The Doctor's occupations, however, were too important to be interrupted by giving instructions to pupils in any branch of science in which he was employing himself for the advancement of his own reputation, and for the benefit of mankind.

He had, for some time past, felt the disadvantage of having neglected classical learning, and had set about, most energetically, to redeem the time he had lost; and from the conviction that a want of knowledge in the learned languages must for ever keep him in the background, he resolved to dedicate his whole time to the study of the Latin and Greek classics, until he had satisfactorily made himself master of them, which was speedily accomplished.

Being now arrived at the age of twenty years, in 1781 he proceeded to Edinburgh, with the view of qualifying himself for the profession of physic, where he took the advantage of attending Dr. Black's lectures. In 1782 he entered as a pensioner of Christ's College, Cambridge, where he studied mathematics, but much more of chemistry and botany. And now he began to exercise his inventive powers, first by a mode of economizing the consumption of fuel in the process of distilla-

tion ; which, however, was not made public until twenty
years after.   He read much, but his rooms were said to
be always in confusion from the mixture of heteroge-
neous materials that were accumulated in them.   His
residence at Cambridge was probably the happiest
period of his life—his spirits unwearied, his health
unbroken, his feelings acute, and his conversation brilliant, though simple and unaffected.

In January, 1785, he was admitted a Fellow of the
Royal Society.   In 1791 he communicated to the
Royal Society his very interesting discovery of a mode
of obtaining carbon from the carbonic acid; and, having
observed that charcoal did not decompose the phos-
phate of lime, he concluded that phosphorus ought to
decompose the carbonate of lime, and the result fully
justified his manner of reasoning.

In 1792 he visited the Continent for the third time,
and, arriving at Paris, perceived an impending convul-
sion, so passed over to Lausanne, where he was highly
interested in the wit and sagacity displayed by Gibbon,
in a conversation which he held with that eminent man.
He next proceeded to Rome and Florence, where
he was fully impressed with admiration of the treasures, of ancient and modern art, possessed by these
two cities; and, returning through Germany, was
greatly amused by the mixture of knowledge and cre-
dulity which he observed among the studious of that
country.

On his arrival in London he took chambers in the
Temple.   In 1796 he took his degree of Doctor of
Physic, and in the same year sent a paper to the Royal
Society on the quantity of carbonic acid afforded by the
diamond, which he measured by heating it with nitre,

and obtaining a precipitate by the addition of muriate of lime; and he found that the diamond afforded no more carbonic acid than an equal weight of charcoal.

In 1799 he gave to the Royal Society a paper on the magnesian limestone, or dolomite, which he considered rather a combination than an accidental mixture. He found that grain, if sown in a soil mixed with carbonate of magnesia, will scarcely germinate, and soon perishes.

In 1802, when making some experiments in crude platina, he discovered in it a singular dark powder, wherein, some time afterwards, he found two new metals to be contained, which he named *iridium* and *osmium*; and for this and various other chemical discoveries he received, in November, 1804, the Copleian medal from the Royal Society.

Soon after this he became more fond of general society, and, somewhat in imitation of Sir Joseph Banks, had his evenings to receive visitors, whom he entertained with curiosities of various kinds; and, in 1812, he was persuaded to convert these mixed exhibitions into a regular course of lectures, calculated for both sexes, and which are said to have highly delighted the whole of his audience. "Their attention," it is said, "was perpetually kept alive by the spirit and variety with which every topic was discussed, by anecdotes and quotations happily introduced, by the ornaments of a powerful but chastened imagination, and, above all, by a peculiar vein of pleasantry, at once original and delicate, with which he would animate and embellish the most unpromising subjects."

In 1813 he delivered a lecture on mineralogy to the Geological Society, in which he introduced an account

of his analysis of a volcanic substance, from the Lipari
Islands, containing the boracic acid.   In the same year
he was elected Professor of Chemistry in the University
of Cambridge.

In 1814 he communicated to the Royal Society a
paper on the easiest mode of procuring potassium; and
another on the economy of heat in distillation, proposing
to heat a second boiler by the condensation of the steam
of the first.   In the spring of this year he was occupied
in searching for the origin of *iodine*, and detected this
substance in sea-water.

One of the last services that he rendered to the Royal
Society was, as a member of a Committee, formed to
investigate, at the desire of Government, the degree of
danger that might attend the general introduction of
gas-lights into the metropolis.   He undertook, together
with his friend Dr. Wollaston, to make some experiments
upon the inflammation of the gas; and they discovered,
conjointly, the important fact, that gas contained in
a small tube will not communicate the flame; a fact
which, in the hands of Sir Humphry Davy, has been
rendered productive of consequences so important to the
public safety; although, Sir Humphry having been
abroad at the time of this discovery, and the report of
the Committee not having then been published, he had
to re-discover this truth, and many more, in his most
ingenious and successful researches.

On the 15th of February, 1815, he arrived at
Calais, where he joined Baron Bulow, went with him
to Boulogne, and thence on horseback to visit Buonaparte's Column: going off the grounds on their return
to look at a small fort of which the drawbridge
wanted a bolt, they were both thrown, with their horses,

into a ditch, by which the skull of Mr. Smithson Tennant was fractured, and he was killed on the spot. Bulow was only stunned.

Mr. Tennant was considered a first-rate chemist, and furnished no less than eight papers, on different subjects, for the ' Philosophical Transactions.' He died in the fifty-fourth year of his age.

Section XI.

Dr. Thomas Young, F.R.S.*

Dr. Young, a man eminently distinguished in more
departments of literature and science than any other
individual of his age and country, was born at Mil-
verton, in Somersetshire, the 13th of June, 1773.   He
might, in fact, be called an Universalist—there was no
art, no profession, no language that he did not compre-
hend.   A sketch of his attainments, from childhood till
he arrived at the age of fourteen, will convey an idea of
his youthful acquirements.

From a very early period he was chiefly an inmate
in the family of his maternal grandfather, Mr. Davies,
of Minehead, who had cultivated a taste for classical
literature, which it was his earnest endeavour to impress
upon the mind of his grandson.   He had learnt to read
with fluency when he was two years old; and it is stated
that, soon after this, in the intervals of his attendance
on a village schoolmistress, he was made to commit to
memory a number of English poems, and even some
Latin ones, the words of which he retained without

* This memoir of the Life of Dr. Young is drawn up, from some
short memoranda of his own writing, in the possession of a near connexion, as stated by one who had the advantage of long and intimate
acquaintance with that distinguished scholar and philosopher, but
who feels himself incompetent to give more than an imperfect
sketch, which he trusts to see filled up hereafter by an abler
hand.

difficulty, although at the time unacquainted with their meaning.

Before he was six years old he attended the seminary of a dissenting minister, and went afterwards to a school at Bristol, where he remained about a year and a half, and where the instructor appears to have advanced very little the studies of his pupil. Young here first became his own teacher, and had by himself studied the last pages of the books used, before the rest had, under the eye of the master, reached the middle. A neighbour of his father, a land-surveyor, indulged him during his holidays with the use of mathematical and philosophical instruments, in his office, together with that of the three volumes of a Dictionary of Arts and Sciences. These were to him sources of instruction and delight of which he seemed never to be weary; and, being thus accidentally thrown in his way, had no small influence on the issues of his future life.

In 1782, then nine years old, he was first sent to the school of Mr. Thompson, at Compton. Here Young went through the ordinary course of Greek and Latin classics, with the elementary parts of the mathematics; and, with the assistance of a schoolfellow, rendered himself tolerably familiar with the French and Italian languages.

He next undertook the study of botany; and, in order to examine the plants, a microscope was wanting, which he attempted to form from the construction given by Benjamin Martin; but for this purpose a lathe was necessary, and the use of this gave way to a passion for turning. Optics followed the microscope, and, falling upon a demonstration in Martin, exhibiting some fluxional symbols, he was never satisfied till he had read and mastered a short introduction to the doctrine of fluxions.

Mr. Thompson had left in his way a Hebrew Bible. He soon read a few chapters, and became absorbed in the study of the Oriental languages. At the age of fourteen, when he quitted Mr. Thompson's school, he was then more or less versed in Greek, Latin, French, Italian, Hebrew, Persic, and Arabic.

When about the age of fourteen, he was attacked by symptoms of what was feared to be incipient consumption; but under the care of Dr. Brocklesby and Baron Dimsdale he recovered his health and was enabled to pursue his labours : during his indisposition he was taking that which, to him, stood in the place of repose—a course of Greek reading. Brocklesby, the eminent physician, and generous friend of Burke and Johnson, was his uncle.

In the four years between 1787 and 1792 he had rendered himself singularly familiar with the great poets and philosophers of antiquity, keeping ample notes of his daily studies. He had acquired a great facility in writing Latin; his Greek verses stood the test of the first scholars of the day; and he read much of the higher mathematics. His amusements were botany and zoology, and to entomology in particular he gave great attention.

In 1790 he attended lectures in chemistry. Dr. Brocklesby soon desired to receive from him a regular report of his literary and scientific pursuits, with a view to his further education for the practice of physic. He communicated some of his Greek translations to Mr. Burke and Mr. Windham, which introduced the young scholar to the acquaintance of those two distinguished persons.

It was probably in these years that his character received its development. He was never known to relax in any object which he had once undertaken, nor during these five years was he ever seen by any one to be ruffled in his temper. He had no faith in peculiar aptitude being implanted by nature for any given pursuit: his maxim was that whatever one man had done, another might do; nor was there anything, he said, that he himself thought worthy to be attempted, which he was not resolved to master.

In 1791 he made his first essays in the press, through the 'Monthly Review' and the 'Gentleman's Magazine:' they consisted of Greek criticism, chemical essays and theories, and remarks on botany and entomology.

In 1792 he pursued his medical studies, attended the lectures of Baillie and Cruickshank, and became a diligent pupil of St. Bartholomew's Hospital.

In 1793 he made a tour into the West of England with a view of studying the mineralogy of Cornwall. He had several flattering offers in that line, but determined to adhere to the pursuit of general science, and to proceed to the practice of physic, as most congenial alike to his predilections and his habits.

In this year he gave to the Royal Society his 'Observations on Vision,' and his 'Theory of the Muscularity of the Crystalline Lens of the Eye,' which became the subject of much discussion, and John Hunter immediately laid claim to his having previously made the discovery. Dr. Young was soon afterwards elected a Fellow of the Royal Society, just as he had completed his twenty-first year.

In 1794 he went to Edinburgh, and there attended the lectures of Drs. Black, Munro, and Gregory, pursuing every branch of study in that university with his accustomed intensity, though he made the physical sciences more peculiarly the objects of his research.

He was extremely fond of music, and of the science of music rendered himself a master. He had at all times great personal activity, and in youth delighted in its exercise. But perhaps it may provoke a smile, though too characteristic an anecdote to omit, that, in instructing himself in a figure of a minuet, he made it the subject of a mathematical diagram.

In 1795 he went to the University of Göttingen and took his doctor's degree. His extraordinary attainments, and wonderful industry, excited the surprise of the laborious school in which he had now placed himself.

In all periods of his life Dr. Young was entirely exempt from those dissipations into which young persons, unhappily, very generally fall; and here, as at Edinburgh, he diversified his graver studies by cultivating skill in bodily exercises. He took lessons in horsemanship, in which he had always great pleasure, and practised all sorts of feats of personal agility, in which he excelled to an extraordinary degree. Before returning to England he visited Dresden and Berlin.

At home he found himself precluded from immediately practising as a licentiate in London, in consequence of some new regulations of the College of Physicians, and therefore entered himself as a fellow-commoner in Emanuel College, Cambridge. He then proceeded to take his regular degrees in physic, living with those most highly gifted, discussing subjects of science with the professors, but not attending any of the public lectures.

His uncle, Dr. Brocklesby, died in 1797 : the larger part of his fortune being left to another nephew, the remainder, with his house, his books, and his pictures, fell to the share of Dr. Young, who now found himself in independent circumstances, surrounded by a circle of academical friends and associates, and moving in distinguished and highly cultivated society, which he continued to enjoy and to prize through life.

His residence at college being completed, Dr. Young settled himself as a London physician in Welbeck Street; where he resided five-and-twenty years. He shortly, however, accepted the situation of Professor of Natural Philosophy in the Royal Institution of London, where he continued two years as lecturer, and colleague with Sir Humphry Davy. He gave two Bakerean lectures on the subject of light and colours to the Royal Society; published a syllabus of a course of lectures on natural and experimental philosophy, with mathematical demonstrations of the most important theorems in mechanics and optics. They contained the first publication of his discovery of the general law of the *interference of light*, being the application of a principle which has since been universally appreciated as one of the greatest discoveries since the time of Newton, and which has subsequently changed the whole face of optical science.

Having married in 1804, he resolved to confine himself mostly to medical pursuits, and to be known to the public solely in that character, but this he found to be impossible. In 1807 he published his course of lectures on natural philosophy and the mechanical arts, in two volumes quarto; the result of an application of five years. It is said to be a mine to which every one has since resorted, and to contain original hints of more things—since claimed as discoveries—than can perhaps be found in any single production of other authors.

Dr. Young did however publish a 'Syllabus of Lectures on the Elements of the Medical Sciences;' and to this he added 'A Preliminary Essay on the Study of Physic,' in which he gives a singular picture of what, in his opinion, is required to constitute a well-educated physician; enumerating every possible quality that a man could wish, but for the attainment of which few could hope.

The 'Quarterly Review' received a variety of articles, literary and scientific, from Dr. Young, many of them connected with the higher departments of science, and containing the results of some of his most laboured researches. The review of Adelung's 'Mithridates' is very remarkable, not only from the immense knowledge it displays of the structure of almost all languages, but as having been the composition which led him to investigate the lost literature of ancient Egypt.

The following year (1814) some fragments of papyri, brought from Egypt, were put into the hands of Dr. Young; and these, with the fragment of the Rosetta stone and the correct copy of its three inscriptions, supplied Dr. Young with materials which enabled him to attach some 'Remarks on Egyptian Papyri, and on the Inscription of Rosetta.'

He now found he had discovered a key to the lost literature of ancient Egypt. He had occupied himself, though without deriving from it the assistance he had at first expected, in the study of the Coptic and Thebaic

version of the Scriptures; but having satisfied himself
of the nature and origin of the enchorial character, he
was content to give the result, anonymously, in the
Museum Criticum of Cambridge.

In 1816 he printed and circulated two letters, relating to his hieroglyphical discoveries and the Inscription of Rosetta, forming the basis on which he
continued his inquiries, as well as the system afterwards carried further in its details by Mr. Champollion,
whose attention had long been directed to similar
studies, in which he has since so greatly distinguished
himself.

In the same year he furnished various articles to the
'Encyclopædia Britannica,' and in this work, under the
head *Egypt*, he first brought out the whole results of his
discoveries in a perfect and concentrated form. In the
same work he supplied sixty-three articles; many of
them biographical, which are admirably given; as are
those likewise on philosophical subjects.

In 1817 Dr. Young had occasion to visit a patient
in Paris, and was greatly pleased with his reception in
the scientific circles, with Messrs. Von Humboldt, Arago,
Cuvier, Biot, and Gay-Lussac.

In 1818 Dr. Young was nominated (as before mentioned, p. 131), one of the paid Commissioners of the
New Board of Longitude, but immediately changed the
character of Commissioner for that of Secretary to the
Board and Superintendant of the 'Nautical Almanac,'
at a salary of £300 a year. Never was there a more
proper appointment, or an office better filled. In the
same year Dr. Young was appointed, by a commission
under the Privy Seal, with Sir Joseph Banks, Sir
George Clark, Mr. Davies Gilbert, Dr. Wollaston,

and Captain Kater, as commissioners for taking into consideration the state of the weights and measures employed throughout Great Britain.

In 1821, after a tour to Italy, and other parts of the Continent, he published an elementary illustration of the 'Celestial Mechanics' of La Place, with some additions relating to the motion of waves and of sound, and to the cohesion of fluids. This volume, and the article 'Tides,' were considered by him to have contained the most fortunate of the results of his mathematical labours. They were spoken of in the highest terms by Mr. Davies Gilbert, from the chair of the Royal Society.

Dr. Young, as a mathematician, was of an elder school, and was possibly somewhat prejudiced against the present prevailing system amongst the continental and the English philosophers; thinking that the powers of intellect exerted by a preceding race of mathematicians were, in no small degree, lost or weakened by the substitution of processes in their nature mechanical.

In 1823 he published his 'Account of some recent Discoveries in Hieroglyphical and Egyptiacal Antiquities,' in which he gave his own original alphabet, his translations from papyri, and the extensions which that alphabet had received from Champollion. At this time he attempted to form a society of about fifty subscribers for the lithography of a collection of plates of Egyptian Antiquities, subservient to the study of hieroglyphical literature. The work was, however, entirely carried on by Dr. Young, and afterwards was made over to the Royal Society of Literature, but continued, during the remainder of his life, to be executed under his supervision.

The study of hieroglyphical literature had taken such strong possession of his mind, that in April, 1829, when under the pressure of severe illness, he could not be persuaded to withdraw his attention from it. He said he had completed all the works on which he had been engaged, with the exception of the rudiments of an Egyptian dictionary, which he had brought near to its completion, and which he was extremely anxious to be able to finish. It was then in the hands of the lithographers; and he not only continued to give directions concerning it, but laboured at it with a pencil when, confined to his bed, he was unable to hold a pen. He said it was no fatigue, but a great amusement to him; that it was a work which, if he should live, it would be gratifying to have finished; and, if otherwise, which seemed most probable, it would still be a satisfaction to him never to have spent an idle day in his life.

His illness continued with some slight variations, but he was gradually sinking into greater and greater weakness till the morning of the following month, the 10th of May, 1829, when he expired without a struggle, having hardly completed his fifty-sixth year. The disease proved to be an ossification of the aorta; not brought on, probably, by the natural course of time, nor even by constitutional formation, but by unwearied and incessant labour of the mind from the earliest days of infancy.

To delineate adequately the character of Dr. Young would require an ability proportionate to his own—one capable of estimating the diversified talents of a man who, as a physician, a linguist, an antiquary, a mathematician, scholar, and philosopher, in their most difficult and abstruse investigations, had added, to almost every

department of human knowledge, that which will
be remembered to after times—" one who," as was
justly observed by Mr. Davies Gilbert, in the before-
mentioned address, "came into the world with a confidence of his own talents, growing out of an expectation
of excellence entertained in common by all his friends,
which expectation was more than realised in the progress of his future life. The multiplied objects which
he pursued were carried to such an extent, that each
might have been supposed to have exclusively occupied
the full power of his mind; knowledge in the abstract,
the most enlarged generalizations and the most minute
and intricate details, were equally affected by him; but
he had most pleasure in that which appeared to be
most difficult of investigation."

To sum up the whole of his character with that which
passes acquirement, Dr. Young was a man in all the
relations of life upright, kindhearted, blameless. His
domestic virtues were as exemplary as his talents were
great. He was entirely free from either envy or jealousy; and the assistance which he gave to others,
engaged in the same line of research with himself, was
constant and unbounded. His morality through life
had been pure, though unostentatious. His religious
sentiments were by himself stated to be liberal, though
orthodox. [He was born and educated a Quaker.] He
had extensively studied the Scriptures, of which the
precepts were deeply impressed upon his mind from
his earliest years; and he evinced the faith which he
professed in an unbending course of usefulness and
rectitude.

Sir Humphry Davy has included among his 'Characters' that of Dr. Young:—

"I must not pass by Dr. Young, called at Cambridge 'Phenomenon Young,' a man of universal erudition and almost universal accomplishments. Had he limited himself to any one department of knowledge, he must have been *first* in that department. But as a mathematician, a scholar, a hieroglyphist, he was eminent; and he knew so much, that it was difficult to say what he did *not* know. He was a most amiable and good-tempered man; too fond, perhaps, of the society of persons of rank for a true philosopher."

The last paragraph, like one of the same spirit directed against Sir Joseph Banks, and which I have already noticed, might, I suspect, be more justly applied to Sir Humphry Davy himself; an intimate intercourse with Dr. Young, as with Sir Joseph, enables me to say that the reproach was in no degree merited by either.

At the conclusion of the little volume from which the foregoing sketch has been taken is 'A Catalogue of the Works and Essays of the late Dr. Young (found in his own handwriting).'

The list enumerates seventy-nine several articles, but affords only very imperfect information, as many of the articles contain a number of volumes, tracts, treatises, &c. As an instance, the articles enumerated under No. 57 amount to about sixty-three.

Section XII.

Sir Francis Chantrey, R.A., F.R.S.

Francis Legott Chantrey, the eminent sculptor, was a Fellow of the Royal Society, and member of the Club ; and though not professing to be a man of science, had a taste for and even a degree of acquaintance with scientific subjects, which qualified him to take a place in both those meetings.   He was born on the 7th of April, 1781, at Norton, a pleasant village about four miles south of Sheffield.   The name of Chantrey had long been known in and about Norton, and it is met with in early and frequent occurrence in the church register.   Their rank in life was humble : one was a huntsman, connected with the family at Norton Hall. The father of Chantrey, from all accounts a very worthy man, was a carpenter, who also rented and cultivated a few fields ; besides these he owned some land distant from the village, the old tenant of which used to tell of the goose-pie which old Dame Chantrey was wont to bring out of the meal-ark on the rent-day.   The farm-cottage which gave birth to our sculptor is said to be still in existence, though somewhat modified ; and so is also the village school at which he learnt to read and write, the only scholastic education it is supposed he ever received.   His father died when he was eight years old, and his mother married again.   One of his earliest occupations, as is stated in the ' Sheffield Mercury,' was to carry milk from the dairy at Norton to Sheffield, in barrels slung on an ass; and that he lingered on the road to make grotesque figures out of the yellow clay—a practice which he is said to have transferred to his mother's butter-closet, where, after churning-days, he was found busily moulding the butter into a variety of shapes and figures, to the great admiration of the dairy-woman.

It is certain that at a very early period of his life he adopted the practice of imitating all kinds of domestic animals in clay; and, once dining with him in after-life, when an ornamented pie came to the table, he said to me, with a hearty laugh, "This same pie brings forcibly to my recollection my having moulded, at the request of a good old dame, for the ornament of her Christmas pie-crust, a sow and pig, taken from the life in her farm-yard: I was then but a boy, but modelling in clay was a passion which daily increased."

As he advanced from boyhood he was placed with a respectable shopkeeper in the grocery line in Sheffield; but it was a business that he could not relish, and begged to be removed from it. It is probable that the master had no objection to release him, on finding that his attention was much more strongly attracted to the shop-window of an opposite tradesman than to his own shop: it was that of a respectable carver and gilder, named Ramsey, to whom, at his own request, he was bound apprentice. At this time Mr. Raphael Smith, mezzo-tinto-engraver and portrait-painter, visited Sheffield in the way of his profession, and, being occasionally at the house of Mr. Ramsey, Chantrey's devotion to the study and practice of drawing and modelling did not escape the artist's observation. Being the first to discover this genius, he took pleasure in giving him instruction.

There also came to the town a statuary of some talent, who taught him as much as he himself knew of the manual and technical arts of modelling, and carving in stone.

His master, no doubt perceiving that his predilection for the arts would render him a less profitable servant, was but little inclined to promote his pursuits. His leisure hours, however, were his own, and were passed chiefly in a small room in the neighbourhood, which he hired for a few pence weekly. Chantrey, however, contrived to separate altogether from Mr. Ramsey before the expiration of his apprenticeship, paying a certain compensation for the remainder of his term. We are told that, being thus free, he visited London, and attended the school at the Royal Academy, but was never admitted as a student. Whence the finances came to defray the expense of this London trip does not appear.

He returned, however, to Sheffield in April, 1802, when only twenty years of age, and put out an advertisement to take portraits in crayons; and in October, 1804, he announced that he had commenced taking models from the life. He also further announced that he " trusts in being happy to produce good and satisfactory likenesses; and that no exertion shall be wanting on his part to render his humble efforts deserving some small share of public patronage." Several specimens of his talent, both in chalk and oil, remain in the town, most of them prized rather for the subsequent celebrity of the artist, than as striking likenesses or as specimens of genius.

He returned to London and was admitted into the Royal Academy, and, having gone through a course of study and improvement, he went back to Sheffield. Here he re-commenced the exercise of his skill in modelling, and executed the busts of five well-known characters, all of which were declared to be such masterly performances, that, when it was resolved to erect a monument to the memory of the Rev. James Wilkinson, Chantrey had the courage to stand forth as a candidate for the commission, though it was said he had never yet lifted a chisel to marble, and it was readily intrusted to him by the Committee. This bold step may be considered as the crisis on which his future fate as a sculptor must be decided. But Chantrey knew well what he was about. He had observed a marble-mason at his work, and soon discovered him to possess superior talents to those of a common workman. Having engaged to employ him in a particular way, he commenced his own task of modelling the monument. The bust being finished, the mason, under the instruction of Chantrey, set about to rough-hew the head for Chantrey's finishing; and this first work of Chantrey may still be seen in Sheffield church.

As London is the great theatre for the display and encouragement of youthful genius, Chantrey resolved to transfer himself, and whatever talent he was master of, permanently to the capital. His first exhibition within the walls of the Royal Academy was in the year 1804, being the bust of a gentleman; in 1805, three busts; and, in 1809, he received an order through Mr. Alexander, the architect, for four colossal busts of four of the greatest and most eminent naval admirals—Howe, St. Vincent, Duncan, and Nelson — for the Trinity Board and for the Greenwich Naval Asylum. His

N

fame was now firmly established, and he purchased a
house and premises in Eccleston Street, Pimlico; and
in the same year married.

In 1810 he executed a bust of Mr. Pitt for the
Trinity House; but 1811 was the year which may be
said to have established his fame and his fortune. In
that year's exhibition he had six busts, all of them pronounced to be most admirable performances:—1, Horne
Tooke; 2, Sir Francis Burdett; 3, J. R. Smith; 4,
Benjamin West, P.R.A.; 5, Admiral Duckworth; 6,
William Baker. An anecdote is told respecting the
conduct of Nollekins, when these busts were setting up
in the Exhibition-room, which does that clever artist
great honour. He lifted one of these busts of Chantrey
(I believe that of Horne Tooke), set it before him,
moved the head to and fro, and, having satisfied himself of its excellence, turned round to those who were
arranging the works for exhibition, and said, "This is
a fine, a very fine *busto*: let the man who made it be
known; remove one of my busts and put this one in its
place, for it well deserves it." Often afterwards, when
desired to model a bust, this same excellent judge
would say, "Go to Chantrey; he's the man for a bust;
he'll make a good bust of you; I always recommend
him." He not only did recommend him, but sat to
Chantrey for his own bust.

It was in this year also (1811) that Chantrey became
the successful candidate for a statue of his Majesty
George III., for the city of London. There is an anecdote told of his being near losing it by mistake. When
the design had been exhibited to, and approved by, the
Common Council, a member objected, having under-

stood that the successful artist was a painter, and therefore most probably not capable of executing, as it ought to be, the work of a sculptor. "You hear this, young man," said Sir William Curtis; "what say you? are you a painter or a sculptor?" "I *live* by sculpture" was the reply—an ambiguous one, it must be confessed; and I cannot guess why he did not give a direct answer to a plain question; but it was accepted, as an assertion that he was a sculptor, and the statue which is now in Guildhall was intrusted to his hands —the first statue that he completed, and it is considered to be very graceful and dignified.

Every year, until 1817, he had demands made upon him, from the highest characters in the nation, for busts or statues. In this year he became an Associate of the Royal Academy; and among the sculptures that were exhibited was that exquisite group of two lovely children, the daughters of the Rev. W. Robinson, of Lichfield, unparalleled in the whole range of modern sculpture. The sisters lie asleep in each other's arms, in the most unconstrained and graceful repose. The snowdrops which the youngest had plucked are undropped from her hand, and both are images of artless beauty, of innocent and unaffected grace. Such was the press to obtain a look at the effigies of these lovely creatures, that there was the utmost difficulty for a great part of the day to get a sight of them: mothers, with tears in their eyes, lingered, went away, and returned; while Canova's now far-famed figures of Hebe and Terpsichore stood almost unnoticed by their side. This exquisite work of art rivalling nature is deposited in Lichfield Cathedral. On a tablet is the following inscription by their widowed mother:—

N 2

"SACRED TO THE MEMORY OF THE TWO ONLY CHILDREN OF
THE LATE REV. W. ROBINSON AND ELLEN JANE HIS WIFE.
THEIR AFFECTIONATE MOTHER,
IN FOND REMEMBRANCE OF THEIR
HEAVEN-LOV'D INNOCENCE,
CONSIGNS THEIR RESEMBLANCES TO THIS SANCTUARY,
IN HUMBLE GRATITUDE
FOR THE GLORIOUS ASSURANCE THAT
'OF SUCH IS THE KINGDOM OF GOD.'"

After the above-mentioned exhibition Chantrey was
overwhelmed with applications, some of which were
irresistible. A bust of John Rennie, the engineer, was
perhaps one of the most striking that ever was made:
it attracted the attention of the whole meeting; parties
crowded round it, but of a different description and
with different feelings from those who pressed forward
to view the sister-children. From the large size of
the head and the fine expressive countenance of John
Rennie, the bust was hailed as a rival to the *Jupiter
Tonans*.

On his return from the Continent he was immediately employed on four busts, which are said to be
among the best where all are good—Lord Castlereagh,
Mr. Phillips (the painter), Mr. Wordsworth, and Sir
Walter Scott—the Wordsworth for Sir George Beau-
mont, and the Walter Scott for his own gratification,
and from mere respect for the worth and genius of Sir
Walter. It is thought that this bust is his very best—
the best, perhaps, in either ancient or modern art—the
man and the genius of the man are both there. It
appears that at first he sought, like Lawrence, for a
poetic expression, and had modelled the head as looking

upward gravely and solemnly. " This," he said to
Mr. Allan Cunningham, when Scott had left after his
second sitting, " this will never do : I shall never be
able to please myself with a perfectly serene expression ;
I must try his conversational look—take him when
about to break out into some sly old story."   As he
said this, he took a string, cut off the head of the clay
model, put it into its present position, and produced, by
a few happy touches, the rudiments of that bust which
preserves for posterity the cast of Scott's expression—
the most fondly remembered by all who ever mingled
in his familiar society.

   This bust, so well known by its casts, is at Abbotsford ; but another of a very similar style and expression
is at Drayton Manor, and never, it seems, has been
copied ; the following letter of Chantrey to Sir Robert
Peel gives an interesting history of both these admirable
busts.

   " *To the Right Hon. Sir Robert Peel, Bart.*

                               " Belgrave Street, 26th January, 1838.
" Dear Sir Robert,
              " I have much pleasure in complying with your
request to note down such facts as remain on my memory concerning the bust of Sir Walter Scott, which
you have done me the honour to place in your collection
at Drayton Manor.

   " My admiration of Scott, as a poet and a man,
induced me, in the year 1820, to ask him to sit to me
for his bust, the only time I ever recollect having asked
a similar favour from any one. He agreed, and I
stipulated that he should breakfast with me always

before his sittings, and never come alone, nor bring more than three friends at once, and that they should all be good talkers. That he fulfilled the latter condition you may guess when I tell you that, on one occasion, he came with Mr. Croker, Mr. Heber, and Lord Lyttelton. The marble bust produced from these sittings was moulded, and about forty-five casts were disposed of among the poet's most ardent admirers. This is all I had to do with the plaster casts. The bust was pirated by the Italians;* and England and Scotland, and even the colonies, were supplied with unpermitted and bad casts to the extent of thousands, in spite of the terror of an Act of Parliament.

" I made a copy in marble from this bust for the Duke of Wellington; it was sent to Apsley House in 1827, and it is the only duplicate of my bust of Sir Walter Scott that I ever executed in marble.

" I now come to your bust of Scott. In the year 1828 I proposed to the poet to present the original marble as an heirloom to Abbotsford, on condition that he would allow me sittings sufficient to finish another marble from the life for my own studio. To this proposal he acceded, and the bust was sent to Abbotsford accordingly, with the following words inscribed on the back:—' This bust of Sir Walter Scott was made in 1820 by Francis Chantrey, and presented by the sculptor to the poet, as a token of esteem, in 1828.'

" In the months of May and June in the same year, 1828, Sir Walter fulfilled his promise; and I furnished from his face the marble bust now at Drayton Manor

---

* Meaning the Italian artists in London.

—a better sanctuary than my studio—else I had not parted with it. The expression is more serious than in the two former busts, and the marks of age more than eight years deeper.

" I have now, I think, stated all that is worthy of remembering about the bust, except that there need be no fear of piracy, for it has never been moulded.

" I have the honour to be, dear Sir,

" Your very sincere and faithful servant,

" F. Chantrey."

That Chantrey had recourse to the process of enlivening his sitter's countenance by the conversation of his friends, I also happened to witness. I dined one day with the late Mr. John Rennie at a *partie carrée* with Scott and Chantrey, while this bust was in progress; and a pleasant dinner we had, for Sir Walter and Rennie were full of *auld lang syne*, and Chantrey and myself delighted—"If I could preserve Scott's look at this moment," whispered the sculptor, "the bust would be perfect;" and with this view he proposed that we four should breakfast next morning in his studio, where the bust and every necessary should be at hand. We went: Chantrey alternately sat down and stood up before his bust, moved it about, kept his eye on Sir Walter, who in telling his own Scotch stories, and listening to those of Rennie, was kept in a state of jocund excitement for, I think, about three hours, when Chantrey, expressing his great satisfaction and delight at the day's proceedings, requested the honour of giving next morning a second breakfast to the same party. All assented; and the same scene was repeated with the same good effect. It seems that Chantrey managed to have relays of friends to meet Sir Walter at his several sittings.

It would require a volume to give a detailed list of the busts, the whole-length statues of single figures and mounted, both in bronze and in marble, and the variety of objects that preserved their form and figure by the operation of his chisel. The two woodcocks which he killed by one shot at Holkham, and carved in marble, may be mentioned as extraordinary specimens of art resembling nature. They were subjects of many poetical effusions, one of which is the following by the Marquis of Wellesley :—

> " Praxiteles, sumptâ pharetrâ telisque Dianæ
>   Venatorque novus per nemus arma movet.
> Acris at illa acies ubi primum intenderet arcum
>   En trajecit aves unâ sagittâ duas.
> ' Parce meis, ne sint vacuæ Latonia sylvis
>   Increpat, et propriâ siste sub arte manum.'
> Ille deæ monitu, atque animosior arte resumtâ,
>   ' Diva,' ait, ' hæc culpæ sit tibi pœna meæ.
> Ponam inter medios, sacrata umbracula, saltus
>   Signa quibus veræ restituentur aves.
> Veræ in morte tamen, quales jacuere sub alta
>   Ilice, jamque anima deficiente pares,
> Aspice! languentes deflexo in marmore pennas!
>   Aspice! quæ plumis gratia morte manet.
> Has tu, Diva, tuas ne dedignare sub aras
>   Accipere, hæc pœnæ stent monumenta meæ.
> Sic tibi lætifico resonet clamore Cithæron,
>   Taygeta, et variis sint tibi plena feris!
> Sic tua delubris auro servetur imago
>   Cui vitam, atque animos, et decus ipse dabo.' "

Besides the numerous pieces of statuary exhibited at the Royal Academy, at least an equal number never entered within its walls. The mention of a few will suffice. Of his statues in bronze there are George IV.,

at Brighton and in Edinburgh ; of Pitt, in the northern
capital and in Hanover Square; of Sir Thomas Munro,
on horseback, at Madras; of George IV., on horseback, in Trafalgar Square; and an equestrian statue of
the Duke of Wellington, for the City of London, in
front of the Royal Exchange.

In 1835 Chantrey received from William IV. the
honour of knighthood.

In 1837 he made a visit to Lord Leicester, at
Holkham, and in returning took Norwich in his way,
to attend the erection of his fine statue of Bishop
Bathurst in that cathedral.

He had left home very unwell, and came back no
better. On the day of his death, which took place
suddenly on the 25th of November, 1841, he looked
over several letters and accounts, gave his orders, and
inspected the progress making in the Wellington
equestrian statue. He was imprudent enough to walk
out in the afternoon, which was raw and foggy, and
returned in great bodily suffering in half an hour. He
ate sparingly at dinner, as his medical attendant had
advised, and said he felt better. Two friends called,
and he desired to see them; they were shown into the
room where he was sitting—but entered only to witness
the last moments of their friend. He fell back in his
chair with a heavy respiration, and expired instantly,
without a word of recognition of his friends. An
inquest pronounced that he died from a spasm of the
heart, which Sir Benjamin Brodie found to have been
the cause; a partial ossification of the heart had taken
place, but the brain was healthy.

Sir Francis Chantrey was one of the most lively,

cheerful, and agreeable men that ever enlivened
society, of which he was fond ; his conversation
abounded with characteristic anecdotes, many of which
I have heard him tell at the Royal Society Club.
Yet it is very remarkable that now, when I am
most desirous of obtaining a sketch of his life,
occupations, and character, I have not been able to
discover any trace of them either in print or manuscript. Yet he was not shy in divulging his pedigree
and the progress of his advancement in life. For
instance, my first meeting with Chantrey was one
Christmas-day at a friend's house where there happened
to be a large ornamented pie on the table, of which I
cannot resist repeating the anecdote I have already told.
Chantrey took the occasion to say, "This Christmas
pie puts me in mind of being asked, when I was a youth
in the country, by an old lady to frame a model of paste
figures to crown her pie. I asked the good lady what
they should be. 'Oh,' she said, 'anything that is
eatable.' A sow with a whole litter of pigs happened
to be running about the yard, and I modelled the whole
group for the lady's piecrust, and when it came to table
it was lauded most unmercifully. And," he added, "I
believe I may say that this group was mainly instru-
mental in making me a sculptor. I was then a mason,
and had not only some practice in cutting stone for
monuments and gravestones, but modelled in clay and
sculptured articles in a coarse way of various kinds.
I was flattered and readily persuaded by the suggestion
of my friends that London was the only theatre on
which I should play the part that was destined for
me; however, I had prudence enough to continue my
humbler practice a very considerable time, and to

bestow much attention on the instruments to be used
in sculpturing.   I was a favourite among the rustics;
they would do anything to oblige me.   If I was bold
enough to endeavour to model a stone head, they
would hew the block into shape, and place themselves
in such positions, and twist their features about, as
I desired.   By perseverance I completed two or three
models, with which I was satisfied, and at last pulled
up courage to make my journey to the capital.   I was
recommended to the gentlemen of the British Museum,
and there I received new and overpowering lights.   I
was most kindly received, and all seemed to be friendly
disposed towards me."

On taking leave that evening, I expressed a wish to
be further acquainted with this agreeable young sculptor,
adding that if I could be of any service to him, I should
be most happy to exert it in his favour.   Many days
had not elapsed before I had a visit from my friend
Mr. Alexandèr, of the British Museum, who said "a
young artist in sculpture had come up with recom-
mendations to the Museum; he tells me that he has
met with you, who have been very civil to him; and it
has occurred to me that, as he appears to be a young
man of good promise, you and I might be of some
service to him.   He has seen Sir Joseph Banks, to
whom he had a letter of introduction; and his remarkable countenance has made so strong an impression on him that he is certain, if he could obtain Sir
Joseph's consent to sit to him, he should succeed in
producing a bust worthy of the original, and one that
would stamp his fortune.   Can you," continued Alexander, "venture to propose such a thing to Sir Joseph?
I know how reluctant he is to be subjected to such

requests, but think, on such an occasion, he would not refuse." " I think with you, and particularly if I propose the request as coming from you and as your recommendation." I did so:

Sir Joseph complied grudgingly; but when the bust was completed and approved, by all who saw it, as a perfect resemblance (which I can testify), and a high work of art, he was quite satisfied; and Chantrey, from that time, had more demands upon his labour than he could comply with.

SECTION XIII.

CHARLES HATCHETT, ESQ.

IF one of our clever novelists, male or female, were to sit down to draw from the stores of their fancy the portrait of a person, who, from the age of boyhood to that of an octogenarian, maintained under all circumstances the same uniform character of fun and frolic, good humour, good sense, kindness, and benevolence, Charles Hatchett might be offered for the original. Nor was he less eminent in the graver walks of literature and science; as a proof of which it may be mentioned that, in 1809, he was elected one of the chosen few of the Literary Club originally instituted by Dr. Johnson and Sir Joshua Reynolds, and on the death of Dr. Burney, in 1829, was appointed to the chief official station of treasurer of the Club. His election into the Royal Society Club is sufficient indication of the estimation in which his scientific, no less than his social qualities, were held by his learned colleagues.

Charles Hatchett was born on the 2nd of January, 1765, at the house in Long Acre where his father carried on the business of an eminent coachmaker. In due time Charles was sent to a school known by the name of Fountayne's, situated in what was called, in former days, Marylebone Park. How long he there continued, and in what manner he finished his education, it does not appear there is any record. " I

only know," says his daughter, Mrs. Brande, " that
he was a tiptop dandy in his day; and one reads an
account of the smart appearance of the boys at Fountayne's school, some described as decked out in pea-
green, others in sky-blue, and a few of the first water
in bright scarlet coats: he, therefore, came home from
school a finished exquisite."

The description given of this fashionable, as it may
be supposed, place of education, would not seem to be
one that was calculated to reconcile the youth to the
coachmaker's business: but I have a curious account of
him in a letter from the Reverend Mr. Lockwood, Rector
of Kingham, who, when he had the living of Chelsea
and Hatchett was residing at his favourite house there,
called Belle Vue, became very intimate, and mutually
dined at each other's houses. " I remember," says
Mr. Lockwood in his note to me, " asking Mr.
Hatchett one day what first led him to turn his attention to the pursuit of chemistry, and he said he believed
it was his love for raspberry jam; for when quite a
boy, accompanying his mother to the storeroom, and
entreating for some jam, she locked the door, and putting
the key in her pocket, jocularly told him he might
now get as much as he could. Somewhat nettled at
this, he set his wits to work; and having read of
the power of certain acids to dissolve metals, he purchased what he thought would suit his purpose, and
applying it to the lock of the cupboard gained an
entrance, and carried off in triumph a pot of jam." If
Hatchett was not hoaxing his friend (which seems very
probable), he may be considered as the first, and will
probably be the last, that a pot of raspberry jam
initiated into chemistry.

There is, however, nothing absurd in the story.
Pope has said—

"What great effects from trivial causes spring !"

In modern prose, we may assert that trivial causes not
more promising than that which unlocked the raspberry-
jam closet, have unlocked a thousand secrets of nature.
Every one at all acquainted with chemical processes
knows how important a part is performed by the
numerous acids; and by Hatchett's own story he had
been reading on the subject of acids, and may we not
easily believe that he had a higher curiosity in testing
the practical power of acids, than merely the gratification of a boyish appetite ?

Whatever may have been the first incentive, there is
abundant proof that he regularly pursued, and with success, his practical experiments on the various subjects
of nature. We find that the first paper of his that
appeared in the 'Philosophical Transactions,' in 1806,
was on a particular *acid*, and is thus entitled :—

'An Analysis of the Carinthian Molybdate of Lead,
with Experiments on the Molybdic Acid ; to which are
added, some Experiments and Observations on the
Decomposition of the Sulphate of Ammonia.'

This was followed by fifteen others on various
subjects, which may here be enumerated, to show that
the extent of his chemical investigations was of no
common description.

1. An analysis of the Carinthian molybdate of lead,
with experiments on the molybdic acid.

2. An analysis of the earthy substance from New
South Wales, called Sydneia.

3. An analysis of the water of the mere of Dis.

4. Experiments and observations on shell and bone.

5. Chemical experiments on zoophytes.

6. Analysis of a mineral substance from North America, containing a metal unknown.

7. Experiments and observations on the various alloys, on the specific gravity, and on the comparative wear of gold.

8. Analysis of a triple sulphuret of lead, antimony, and copper from Cornwall.

9. Analytical experiments and observations on lac.

10. Analysis of the magnetical pyrites; with remarks on some of the other sulphurets of iron.

11. Observations on the change of some of the proximate principles of vegetables into bitumen; and experiments on a peculiar substance found with the Bovey coal.

12. On an artificial substance which possesses the principal characteristic properties of tannin.

13. Additional experiments on the above subject.

14. A third series of experiments on the same subject.

15. A description of a process by which corn tainted with must may be completely purified.

His election into the Royal Society was on the 9th of March 1797; his age thirty-two years.

Mr. Hatchett, it appears, was intended to follow his father's business, and lived with him on his removal to excellent premises in Hammersmith; but the young man never took kindly to it, a great portion of his time being employed on books of science, and particularly those which treated on chemistry. He attended lectures on scientific subjects; and his father,

on perceiving the bent of his inclination, made him
a handsome allowance to enable him to prosecute his
studies, even to his own disadvantage; for on a pressing
occasion, when obliged to supply his son with a little
money, he was compelled to dispose of a magnificent
collection of minerals, to part with which gave him no
little annoyance.

The occasion, however, appears to have been important. On the 24th of March, 1786, when young
Hatchett was just one-and-twenty, he married the
only daughter of Mr. John Collick of St. Martin's
Lane, a man of good property, engaged in the extensive
business of a hair-merchant: wigs being then almost
universally worn, the trade in hair was a lucrative
concern. The young couple, of the same age, appeared
so very juvenile, that the lady of the boarding-house
at Bath, where they passed the honeymoon, set it down
as a runaway match. They visited Russia and Poland,
where they remained about two years. During their
stay, Hatchett made acquaintance with several distinguished persons of all classes; was introduced to
Stanislaus, King of Poland, and many Polish noble-
men and ladies. He dined in company with the
Chevalier d'Eon, who was there disguised as a lady;
and after dinner retired with the ladies, to the no small
discomfiture of those who were in the secret.

On their return they lived at Hammersmith, where
in a short time his mother died, to his great grief, for he
loved her dearly; and he always said that he was indebted to his mother for his great love of reading. He
established himself in a good house in Hammersmith,
where he got up an excellent laboratory, and collected
a good library. " My father," says Mrs. Brande, "was

o

exceedingly musical, and played well on the piano, but much better on the organ. He had been a pupil of Dr. Cooke's, and was acquainted with the first musicians of his day; as was also my mother, considered to be one of Clementi's best pupils." Hatchett was not an useless or inactive Fellow of the Royal Society; on the contrary, his assistance was frequently called for.

In 1810, Charles Hatchett, assisted by Drs. Wollaston and Babington, with four other gentlemen, were appointed to value the minerals of Mr. Greville, in pursuance of a petition to the House of Commons from the Trustees of the British Museum. They reported that the whole collection consisted of about 20,000 specimens; that the series of crystallized rubies, sapphires, emeralds, topazes, rubetites, diamonds, and precious stones in general, as well as that of various ores, far surpasses any known to them in the different European collections; and that the value of the whole was 13,727l., including that of the cabinets, which cost 1600l.

Hatchett was one of the chemists consisting, at a meeting at Sir Joseph Banks's, of Brande, Hatchett, Wollaston, and Young, by whom a resolution was made—that Mr. Stephenson was not the author, as certain persons had asserted, of the discovery of the fact that an explosion of inflammable gas will not pass through tubes and apertures of small dimensions; and that Mr. Stephenson was not the first to apply that principle to the construction of a safety-lamp: but that Sir Humphry Davy had the sole merit of having first applied that principle of the non-communication of explosions through small apertures, the discovery having been made by the experiments of Mr. Tennant on flame.

In 1818 Charles Hatchett and Dr. Hyde Wollaston were joined in a commission which passed the Great Seal, appointing and authorising an inquiry into the best means of preventing the forgery of bank notes. With these two practical men of science were joined Sir Joseph Banks, Sir William Congreve, Davies Gilbert, Mr. Hurman (Governor of the Bank) and some others.

Hatchett was, in fact, very commonly called upon committees, whenever points of chemistry or other sciences were to be discussed, but they never interfered with his familiar tone of jocularity when anything occurred that struck his fancy. At the Club he was always most agreeable, and was a great favourite with every one, particularly so with Sir Joseph Banks, because he also was exactly what Sir Joseph, after Dr. Johnson, called a clubbable man; he was good-humoured, full of drollery, never at fault for some jocular or pleasant story to amuse the company. "Hatchett," he would say, "is an excellent chemist, and abounds in good solid sense on most points, and he is not spared in the business of the Society, where we can never have too much of science and philosophy; but I consider a social meeting like this of the Royal Society a relaxation from the more laborious studies." Sir Joseph being President of the Club as well as of the Society, when a new member was proposed always asked, "Is he a clubbable person?"

One day at the Club, Hatchett amused us with the story of a dream which he had had, and which he prefaced by saying it was "such stuff as dreams are made of," but that it contained a reality in its conclusion which had very much distressed him. He dreamed that he had lost his way, but came to a dark dismal-looking

o 2

building, into which he passed through a forbidding sort
of a gate, opened to him by a black-looking porter, who
closed it immediately upon him. He walked on, and
everywhere observed clumps of ill-looking people skir-
mishing and fighting; a little beyond, other groups were
weeping, and wailing, and gnashing of teeth ; farther on
still were flames of fire; and, beginning to think that he
had got into a very bad place, he endeavoured to retrace
his steps and get out again, but the black doorkeeper
refused to let him pass. A furious fight ensued, and
he pummelled the negro-looking rascal first with one
fist then with another, when he was brought to his
senses by a female scream, which, to his dismay, proceeded from his poor wife, and he found that, instead
of pummelling the black doorkeeper, he had given Mrs.
Hatchett a black eye!

In 1810 Hatchett had taken up his residence at
Belle Vue House, in Chelsea, an abode in which he continued for the remainder of his life : he prosecuted his
studies, and entertained his friends and neighbours with
music, books, pictures, and a variety of curiosities.
Mr. Lockwood, who has already been mentioned, thus
speaks of Hatchett :—

"During the four years I resided at Chelsea I was
constantly in the habit of seeing Mr. Hatchett; he used
every now and then to come and sit with me of a morning for half an hour, and nothing could be more agreeable than his conversation. I used also often to visit
him in his library, which he always called his *den*. It
was a large room, well supplied with valuable books, and
contained an organ, for he was very fond of music. He
was very hospitable, and rather prided himself on
the excellence and variety of his wines. I met at his

table men whose talents had rendered them conspicuous
in the world; but he was himself always the life of the
party, with his anecdotes and jokes, and very pleasant
his parties were : among his most constant guests were
Sir Francis and Lady Chantrey, and Dr. and Mrs. So-
merville.

"There is a print of him which is very like; and
when he gave me a copy, he repeated the lines written
by Jekyll, on receiving a similar present, on the 30th
of January, the martyrdom of Charles."

The *lines*, if *verses* be meant by Mr. Lockwood, were
not Jekyll's, but James Smith's, the author of the
'Rejected Addresses;' though the original joke was
Jekyll's, in a prose note more sharp and terse than the
subsequent versification. The occasion was this:—On
the 29th of January Hatchett sent a portrait of himself
to his friend Jekyll, and on the following day, 30th of
January, received the following note :—

                                "January 30, 1846.

"Thanks for a kind memorial of our long friendship, though it looks somewhat 'radical' to thank the
Hatchett for the head of Charles.

                                "JOSEPH JEKYLL."

This James Smith thus rhymed :—

> "An answer, Charles Hatchett, thou claimest,
>   So take it, both pithy and short ;
> For surely so able a chemist
>   Can never reject a retort.
> Your portrait no painter can match it,
>   So I scorn all their envy and snarls ;
> And, like Cromwell, I owe to a Hatchett
>   What I gain by the head of a Charles."

I have said that Mr. Hatchett was a member of the

celebrated Club founded by Dr. Johnson and Sir Joshua
Reynolds, known as the Literary Club, and, by themselves, as *The Club, par excellence.*   Mr. Croker,
wishing to give a complete list of the members of the
Club, applied to Mr. Hatchett, whose answer he inserts
(in vol. i. p. 522 of his ' Boswell's Life of Johnson,'
Ed. 1832) with this note :—"The following complete
list of *the Club*, with the dates of the elections of all
the members, and the deaths of those deceased, from
its foundation to the present time, and the observations prefixed and annexed, have been obligingly
furnished to the Editor by Mr. Hatchett, the present
Treasurer."

Hatchett not only supplied him with the names of the
nine original founders, the whole list of the elected
members, and of those who had died, but he also
furnished the following history of the Club :—

"The Club was founded in 1764 by Sir Joshua
Reynolds and Dr. Samuel Johnson, and for some years
met on Monday evenings.   In 1772 the day of meeting
was changed to Friday; and about that time, instead
of supping, they agreed to dine together once in every
fortnight during the sitting of Parliament.

"In 1773, the Club, which soon after its foundation
consisted of twelve members, was enlarged to twenty ;
March 11, 1777, to twenty-six; November, 1778, to
thirty; May 9, 1780, to thirty five ; and it was then
resolved that it never should exceed forty.

"It met originally at the Turk's Head, in Gerard
Street, and continued to meet there till 1783, when
their landlord died, and the house was soon afterwards
shut up.   They then removed to Prince's, in Sackville
Street; and, on his house being soon afterwards shut up,

they removed to Baxter's, which afterwards became Thomas's, in Dover Street. In January, 1792, they removed to Parsloe's, in St. James's Street; and, on February 26, 1799, to the Thatched House, in the same street.

"From the original foundation to this time the total number of members is one hundred and two. *Esto perpetua.*

"C. H.
"Belle-Vue House, Chelsea, July 10th, 1829."

Mr. Hatchett's daughter, Mrs. Brande, has kindly sent me a few extracts from his occasional correspondence, which, she observes, speaks more for his natural talent for pleasantry than for any grave and scientific remarks. His first letter expresses a hope of having got rid of every symptom of his late attack of giddiness, which occasioned him a very alarming fall down stairs, and obliged him to be cupped. "In fact," says his daughter, "it was an attack of paralysis, from which in time he quite recovered, but he gives no thanks to the doctors." "I really think," he says, "that the mode of treatment in such cases as mine is as bad, or worse, than the disease. To eat little —wine not to taste, or in small quantities—not to look up, because it may produce giddiness—not to stoop, nor look down, for the same reason. I drop my handkerchief, and am going to stoop for it—I recollect the prohibition—I look at the handkerchief, and am inclined to give it a left-handed blessing; but then, I have been particularly cautioned not to get into a passion. I must not take snuff, because it may produce sneezing—I must not go out in the carriage, it may cause giddiness—and not think much, nor read much, —not write much, nor talk much. I ought to avoid laughter, because it acts upon the diaphragm and drives the blood to the head. I have been seriously advised not to crack more jokes than I can help, and therefore to admit as few friends as possible. The practical part of music may be injurious, because it is exciting. In short, the mode of cure may thus be summed up— the patient is to abjure every comfort, whether corporeal or mental, so as to become as nearly as possible a log, with the miserable consciousness of being alive! This is a fine specimen how attentive I am to the prohibition of ' not writing much ;' and, therefore, my dear Frederica, I will conclude with my love to all."

The fact seems that Hatchett, like many, perhaps most of us, loved to doctor himself. This would appear from another letter, where he says, — " My medical man accidentally called the other day, and advised me to take but little wine *after dinner* ; and as a substitute for wine *at dinner*, he recommended barley-water with gum-arabic. I could not, of course, listen to this mode of proceeding, which I assured him I had not employed during the last fifty years."

Again he says,—" After having had the advice of two very eminent surgeons without any very beneficial effect, I consulted myself, and by animal chemistry have successfully eradicated the rascally corn which had crippled me for the last three months."

Alluding to mesmerism, he says,—" There always has been, and always will be, some medical foolery. When Dr. Mesmer displayed his wonders in Paris, they excited so much attention, that, by order of

Government, a committee was formed of *Messieurs de
l'Académie des Sciences* (the celebrated Dr. Franklin
being one of them), with the intent of ascertaining
whether the effects had any real foundation in scientific
truth. The report of those gentlemen (which I have)
was printed, and stated that the whole was deceptive,
and resulted from excited imagination and irritation of
the nerves in credulous persons. *I* was mesmerised,
and so was *John Hunter*; but we were both pronounced
to be *bad subjects.*"

When Mr. Hatchett was in Norfolk he fell in with
Mr. Dawson Turner, who informed him that he had
some idea of writing the Life of Sir Joseph Banks, and
asked him if he could supply him with any particulars
and anecdotes which might occur to him. "Since
then," says Hatchett, "I have amused myself by composing what may be called *Banksiana*, and have really
got together a pretty set of stories, many of them
being highly seasoned, and, of course, very edifying."
Whether they were ever printed, or what became of
them, I know not.—I have no trace of them; and I
suppose that they were mere scattered memoranda,
which there is little chance of recovering; but should
they ever come to light, there can be no doubt that the
sayings and doings of Sir Joseph Banks, recorded by
Charles Hatchett—a warm admirer and great favourite
of Sir Joseph's—will furnish a great treat.

Hatchett's taste for reading, his daughter says, continued to the last; and alluding to the Diary of Madame
d'Arblay he writes, " The review of the 'Quarterly' is
certainly very severe, although, in my opinion, fully
merited; and yet I have read her work with much

amusement and interest, for many of the persons mentioned by her were known to me in my boyhood and early youth. Often, however, she merely mentions them, and of course causes disappointment to those who were acquainted with them. I knew Piozzi; have heard him sing, and have some canzonets of his composing which he gave to me; they are somewhat elegant, but flimsy. He was a handsome, dark Italian, of interesting manners, and likely to captivate a widow of sixty—a dangerous age; for they are then like *dry old stubble*, soon in a blaze."

He says in another of his letters, " I grow every day more and more saucy, so as even to expect the widows and spinsters to come to me, instead of my going after them: and as to domestic conversation, it is quite outrageous for wives to expect husbands to come home brimful of news, anecdotes, and fun. No, no; the wives must entertain them, or contrive to stimulate in some way or other." When the Bishop of Durham (Barrington) told Archdeacon Paley that he and Mrs. Barrington never had had a single word of difference in the course of thirty years, Dr. Paley observed, " Indeed, my lord! very flat!" And Hatchett adds, " Some of those little breezes quicken matters, like a squeeze of lemon in punch, and remove the ennui expressed in the old French chanson, by the refrain, 'Le triste badinage de l'amour conjugal.' "

In one of his letters he tells an anecdote which was talked of at the time: whether it was he or his friend Jekyll, or neither, who gave it currency, he does not tell us. It is, at least, founded on fact. " At the time that the late Lord Stowell married the Dowager

Marchioness of Sligo, he was then Sir William Scott, and the plate on his door exhibited

> Sir William Scott.
> Marchioness of Sligo.

on which Jekyll observed, that Sir William was *knocked up*. This joke got about, and the plate was speedily altered to

> Marchioness of Sligo.
> Sir William Scott.

"Now," said Jekyll, "Sir William knocks under."

In a letter of the 21st of May, 1845, he relates to Mrs. Brande the great pleasure he had received from a visit of a member of the Royal Society Club, to entreat him to withdraw his resignation.

" Yesterday Captain Smyth called on me, having been deputed by the Royal Society Club earnestly to solicit that I would not withdraw my name from the Club, and that I would continue to be in my place as senior member or Father of the Club. This was expressed in the most handsome and kindest complimentary language, with an assurance that all outstanding laws and regulations of the Club should be waived, in case that I would consent that my name should remain. It was impossible for me to refuse my compliance with such an affectionate request, and I therefore gave my consent in terms expressing how warmly I felt the honour done to me by the Club."

To this Mrs. Brande adds, "My father, being at this time in his eighty-first year, must have been a member of the Royal Society Club many, many years, and naturally felt proud of the kind solicitations of the then existing members.   In another letter of the same year, my father, not being well, and not caring for country retirement, says, 'After all, I begin to believe, like my old friend Mr. Jekyll, that there is no place better than the metropolis, nor any rurality equal to the Parks, where, as Jekyll did, I join the truly gay promenade almost every afternoon about three o'clock, and dowager in my carriage backwards and forwards along the banks of the Serpentine.   One of the constant promenaders, Lady Holland, will not, I fear, enjoy many more, for she is dangerously indisposed.   Almost all my old friends and acquaintances are dead, or something like it ; for the few survivors become old, and very dull and stupid.'"

"This," she adds, "was the last letter I received from my dear father: it was about the commencement of his last illness—water in the chest.   He died on the 10th of March, 1847, sensible to the last."

And I, who survive to inherit his venerable station, as Father of the Royal Society Club, am writing this on the 29th day of October, 1848, in the eighty-fifth year of my age.

POSTSCRIPT.

Within a month of writing the last paragraph*—that is, about one o'clock P.M. *on the 23rd November, 1848—and three days after the decease of his son-in-law, Lieut.-Colonel Batty—Sir John Barrow's long, useful, and honourable life was closed, by a death so sudden, and yet so placid and happy, as to be worth recording. He had during the forenoon of that day walked out, as was his daily custom, and had transacted some business at a public office, with his usual activity of body and clearness of mind, and seemed, to those who met him, as well as ever he had been in his life ; on returning home, he sat down to lunch with his family, and after having asked for something on the table, before he could partake of it—without a struggle or even a groan—he ceased to live—by, as is supposed, an instantaneous cessation of the action of the heart.*

The following paragraphs of the Ulverston Advertiser *of the 28th December, announced this event to the neighbourhood in which Sir John was born, and which he had always regarded with particular interest.*

. . . . . . " Sir John Barrow never forgot the spot that gave him birth. By his will, the annual subscription which he had been in the habit of contributing for a long series of years to the support of the School in which

* Which has been selected to give a specimen of his handwriting, which, though less firm than his early character, is still remarkable for a man of eighty-five, without glasses.

he was educated, is to be continued, and his cottage at Dragleybeck given over in perpetuity to trustees, that the rent may be appropriated to the education of the poor at the same school.

" His memory will long survive, and his example be held up for imitation by all who derive their birth or education from the same locality.   The name of *Sir John Barrow* is a household word amongst us. Although he who bore it is departed, his memory still lingers lovingly about our hearths, and will continue to be cherished by our children's children, through many a generation.

" Sir John having expressly desired that his Funeral should be quite private, none will be expected to attend it beside his three sons and grandson, and his old friends, Sir George Staunton, Sir Benjamin Brodie, and the Right Honourable John Wilson Croker, his former colleague and near connexion through the marriage of the present Sir George Barrow to a sister of Mrs. Croker. Yesterday, Wednesday, 29th November, being the day of the interment, it was observed at Ulverston by the tolling of the bells of the old church ; and a blue ensign, half pole high, waved over the cottage in which he was born !"

The following Summary of the Life and Character of Sir John Barrow appeared in The Times *a few days after his decease, from the pen of his early and constant friend, Sir George Staunton :—*

"The name of Sir JOHN BARROW will occupy an honourable place in the list of those highly gifted individuals of whom England is justly proud, and who, by their original genius and energetic minds, have, in their different walks of life, rendered eminent services to their country. The friends of his childhood and youth did not provide him with more than the ordinary means of instruction, but he seized on those means with avidity and industry, and it was his self-education that mainly conferred on him those powers which, when the day of trial arrived, he turned to so good an account.

" About the time that Mr. Barrow arrived at the period of manhood, he was fortunate in obtaining, through the interest of a friend, a place in the first British Embassy to China. He was thus enabled to put his foot on the first step of the ladder of ambition; but every subsequent step of his advancement in his distinguished career may be fairly said to have been achieved by himself. His talents and his zeal for the public service, when once known and placed in a fair field for action, could hardly fail of being appreciated and duly fostered by those distinguished statesmen under whom he successively served.

" It so happened, that the chiefs of the British Mission to China in 1792, the Earl of Macartney and the late Sir George Staunton, were, in some respects, not so happily provided with active and talented associates as might have been wished: but in Mr. Alexander, the draughtsman of the embassy, they were fortunate in possessing a very able and diligent artist; and Mr. Barrow, from his various talents, and the zeal and alacrity with which he applied himself to every department of the service, although his own was only a subordinate one, was a host in himself. The authentic account of the embassy, published by the late Sir George

Staunton, records many of Mr. Barrow's valuable contributions to literature and science connected with China. This work, therefore, together with his own subsequently published supplemental volume of travels, is ample evidence how well his time had been employed. Had no unpropitious political events occurred to prevent the views and plans of the mission being carried out, it is not too much to say that the able and ingenious men who were employed in it would most probably have effected, by peaceful means, all those improvements in the terms of our intercourse with China which, some fifty years after, have cost us such a painful expenditure of blood and treasure. It was not to be expected that any person of mature age could within the space of a few months overcome all the practical difficulties of such a language as the Chinese; but Mr. Barrow had already begun to converse in it, and he had acquired a complete knowledge of its theory. His papers on this subject in the *Quarterly Review* contain probably the best and most popular account of that singular language and character which was ever presented to the British public.

" Although Mr. Barrow ceased to be personally connected with our affairs in China after the return of the embassy in 1794, he always continued to take a lively interest in the varying circumstances of our relations with that empire. On the occasion of the second embassy, under Lord Amherst, in 1816, he was of course consulted by the ruling powers; but, unfortunately, although his advice was asked, it was not taken; and in consequence of the injudicious rejection of the proposal which his prophetic sagacity had suggested for getting rid of the vexatious question of the Chinese ceremony, Lord Amherst and his colleagues were compelled to abandon the personal reception of the mission for the sake of preserving the honour and real interests of the English in China, which would have been essentially damaged by the acceptance of the terms upon which it was offered. Mr. Barrow was likewise consulted, and we believe more fairly and confidentially, on the occasion of our recent conflict with China, which, it is to be hoped, has secured our future peace with that country.

" Lord Macartney was naturally anxious to secure the aid of such a man as Mr. Barrow in his next public service, his important and delicate mission to settle the government of our newly acquired colony of the Cape of Good Hope. Mr. Barrow was intrusted with our first communication with the Caffre tribes, and it would have been well for the public interests if the spirit, judgment, and humanity which he then displayed, had more uniformly governed our subsequent transactions with that remarkable race. The two volumes of his History of the Colony, and the unrivalled map with which they are illustrated, made the public at once fully acquainted with the extent, capacities, and resources of that important, but till then little understood acquisition of the British Crown.

" There is little doubt that it was the perusal of this valuable work which mainly decided Lord Melville to accept of Lord Macartney's recommendation of a perfect stranger to him, as Mr. Barrow then was, as his Second Secretary of the Admiralty. It is not our purpose to enter here into the merits of Mr. Barrow's subsequent career for forty years at the Admiralty. It would be, in fact, nothing less than the history of the civil administration of our Navy for the same period. Suffice it to say, that he enjoyed the uniform esteem and con-

fidence of the eleven Chief Lords who successively presided at the Admiralty Board during that period, and
more especially of William IV., while Lord High Admiral, who honoured him with tokens of his sincere
personal regard.   Mr. Barrow received the honour of
the baronetcy during the short Administration of Sir
Robert Peel in 1835; and strong as party feeling ran
at that time, not a voice was heard in disapproval of this
exercise of the Royal prerogative.

" He no doubt held strong opinions on the various
national questions upon which the great political parties
in this country are divided,—as who does not who has
a heart devoted to his country, and a head capable of
serving it!   But he never suffered any party feeling or
bias to interfere with the zealous discharge of his public
and official duty; and it so happened, that the most
remarkable and active period of his public service at the
Admiralty was that during which he was occupied in
carrying out those important changes, which have so
much improved and simplified the system of the civil
administration of the Navy, and which were introduced
by Sir James Graham, under a Whig Administration.

" Sir John Barrow retired from public life in 1845,
in consideration of his advanced years, although he was
still in vigorous possession of all the mental and bodily
powers required for the due discharge of the functions
of his office.   In the course of the succeeding three
years his vital energies must no doubt have become
gradually somewhat weaker; but he seemed on the whole
so hearty and so fully in the enjoyment of his faculties,
that his friends and relatives observed nothing like
decay, and could have entertained no apprehension that
his end was so near.

" Sir John Barrow had the moral courage to publish, during his life-time, his own biography, and he modestly states his motives in the following words:—' To trace my progress through the vicissitudes of a life extended beyond the general period of human existence, and, by the mercy of God, without any painful suffering from accident or disease, has been my object, more with a view of benefiting my children and theirs, by the example it holds forth of industrious habits, than with any other.'—Page 488. We are sure the public have been thankful to him for this interesting addition to his already numerous publications, and will wish that other eminent men, whose career has been similarly distinguished, and similarly worthy of imitation, may follow his example.

" We have said nothing yet of his various other works, whether published in his own name, or anonymously inserted in various Reviews, chiefly the *Quarterly*; because they are already well known to the public, and speak for themselves. During a long series of years, whenever an article illustrative of science or enterprise appeared in the *Quarterly Review*, the public at once recognised the hand from which it proceeded, and valued it accordingly. He had, indeed, not only a remarkable facility in composition, but, what was of still more importance, that of detecting, sifting, arranging, and applying all those dispersed and often obscure materials which were essential to the elucidation of his subject, but which, however important in themselves, had been, in their crude state, almost unknown and valueless. He was, however, surprised, when his publisher, Mr. Murray, presented him with ten portly and handsomely bound volumes, containing the Essays of his own composition, selected from the *Quarterly*, and comprising at least one-fourth part of that periodical, as it then existed.

" It it impossible, even in this brief Memoir, to pass over altogether without notice one remarkable feature of Sir John Barrow's official life,—his advocacy and promotion of the several Polar Expeditions.　Although it is absurd to impute the direct responsibility for these expeditions to any other quarter than the several Administrations during which they were undertaken, there can be no question but that these enterprises originated in Sir John Barrow's able and zealous exhibition, to our Naval Authorities, of the several facts and arguments upon which they might best be justified and prosecuted as national objects.　The anxiety just now beginning to prevail respecting the fate of Sir John Franklin and his gallant companions, throws at this moment perhaps somewhat of a gloom on this subject; but it ought to be remembered that, up to the present period, our successive Polar Voyages have, without exception, given occupation to the energies and gallantry of British seamen, and have extended the realms of magnetic and general science, at an expense of lives and money quite insignificant, compared with the ordinary dangers and casualties of such expeditions, and that it must be a very narrow spirit and view of the subject which can raise the cry of ' Cui bono,' and counsel us to relinquish the honour and perils of such enterprises to Russia and the United States of America!"

30 *November*, 1848.

London : Printed by WILLIAM CLOWES and SONS, Stamford Street.

Printed in the United States
By Bookmasters